T0226104

Technology Standard of Pipe Rehabilitation

Lu Wang · Chunwen Yan · Junyu Xu

Technology Standard of Pipe Rehabilitation

CHINA ARCHITECTURE & BUILDING PRESS

Lu Wang
School of Engineering and Technology
China University of Geosciences (Beijing)
Beijing, China

Chunwen Yan
China Geological Equipment
Group Co., Ltd.
Beijing, China

Junyu Xu
China Society of Trenchless Technology
Beijing, China

ISBN 978-981-33-4986-5   ISBN 978-981-33-4984-1   (eBook)
https://doi.org/10.1007/978-981-33-4984-1

Jointly published with China Architecture & Building Press
The print edition is not for sale in China (Mainland). Customers from China (Mainland) please order the
print book from: China Architecture & Building Press.
ISBN of the Co-Publisher's edition: 978-7-112-19471-1

This Springer imprint is published by the registered company Springer Nature Singapore Pte Ltd.
The registered company address is: 152 Beach Road, #21-01/04 Gateway East, Singapore 189721,
Singapore

Preface

Underground pipelines are the critical infrastructure of the city. They include the water supply and drainage system, gas pipeline, electric power pipeline and communication pipeline, transporting energy and removing waste for residents, factories and enterprises located in different places of the city. The pipes for the city likes the blood vessels for the human body. Therefore, they are called as the "lifeline of the city." However, with the development of the city, due to years of use, underground pipelines are tending to be aging and have some defects,such as cracks and leaks. If they are not repaired or renewed in time, the city operation will be seriously affected.

Trenchless pipe rehabilitation technology is a new technology that could repair and renew the original pipeline without excavating the ground. It does not affect the normal life around the pipeline. What is more, it can save the construction time. As a result, it lowers the cost compared to the traditional rehabilitation method. Therefore, it is not only very welcomed by the trenchless construction units, but also supported by the government. Thus, it has been developed rapidly in the recent ten years.

However, due to the trenchless pipe rehabilitation technology is new and is not developed maturely, the pipeline renovation construction by this technology is chaotic and do not have standard to reference. The relevant terminology are not unified which is not good for communication. In this situation, authors compiled this technical regulation which is referring to various international advanced standards and relevant theories, summarizing domestic and foreign experience of trenchless pipe rehabilitation projects and combining personal construction experience for many years.

This standard summarizes the technical method and construction process of underground pipeline testing, cleaning and renovation. It has 20 chapters and two appendixes in total. Its content includes: Pipeline rehabilitation construction organization design, pipeline cleaning, preparations before construction, pipeline detection and quality assessment, pipeline renovation design/method/equipment selection/steps/technical indicators, universal construction techniques, construction general rules, the engineering quality acceptance, construction health, safety, environmental protection and production management, and so on. This book introduced these pipeline renovation methods in details: Pipe cracking, sliplining, pipe segmental lining, modified sliplining, cured in place pipe (CIPP), spray lining, spirally wound lining, localized repair. The first appendix is the interpretation for the relevant technical terms in this book. The second appendix is pipe defect grade classification and sample figure. It could help the reader who do not have the basic knowledge about pipe rehabilitation to understand this technology easily.

This standard could be the fundamental discipline for pipeline renewal projects in different industries. It could provide the important basis and criterion for design, construction, management, inspection and acceptance of pipeline renewal projects. It can be used by designers, managers, researchers, engineers and technicians from municipal pipeline engineering, petroleum pipeline engineering, water supply and drainage engineering, environmental engineering, geological engineering and other fields, can also be used as the teaching material for teachers and students.

The original version of this book was proposed by the pipeline renewal expert group of trenchless technical Committee of Chinese Geological Society and relieved in Chinese in 2016. Since its publication, it has been used as the teaching materials for trenchless technology training. All the readers who read this book think it is very helpful. This book has been the bestseller in China. A lot of foreigners want to read this book, but they do not know the Chinese characters. With the support and authorization of the authors who are the original version, we translated this book into English. Hopefully, this book could benefit all the people all over the world.

Because of various reasons, some contents have not been included in this standard, such as: pipe eating, pipe pulling, engineering accident prevention and treatment, design optimization and so on.

Besides, due to different pipeline rehabilitation construction purposes, some technical indicators and requirements may have been changed. In this case, the pipe construction process should comply with the specific provisions of the agreement and contract. Pipeline cleaning and testing, engineering design, construction and quality acceptance of pipeline renewal projects shall not only comply with this standard, but also should comply with the mandatory provisions of the current relevant standards issued by the state and relevant underground pipeline authorities.

Beijing, China   Lu Wang
                                                                                   Chunwen Yan
                                                                                    Junyu Xu

Acknowledgments

The author(s) disclosed receipt of the following financial support for the research, authorship and/or publication of this book: This work was supported by the Natural Science Foundation of China (41902320) and the Fundamental Research Funds for the Central Universities (2652018089).

Thanks to Prof. Wenjian Zhu of Beijing Institute of Exploration Engineering for offering professional advice for this book. Thanks to Senior Engineer Fangjun Li and Mingqi Wang for supporting the translation work of this book. Thanks to Prof. Yuanbiao Hu of China University of Geoscience (Beijing) for revising and proofreading the English of this book. Thanks to my students Pingfei Li and Xuesong Bai for helping me search and organize the materials. In addition, the authors express their appreciation to China Society for Trenchless Technology for providing data and materials.

Contents

Chapter 1
Scope

This standard specifies the technical requirements and relevant management standards for all procedures in the construction of trenchless pipeline rehabilitation (replacement and renovation).

This standard is suitable for the construction project which utilizing the trenchless pipeline rehabilitation technology to renovate and upgrade the underground pipelines.

© China Architecture & Building Press 2021
L. Wang et al., *Technology Standard of Pipe Rehabilitation*,
https://doi.org/10.1007/978-981-33-4984-1_1

Chapter 2
Normative Reference Documents

The following documents are necessary when you are using this standard. For any dated references, only the dated version applies to this standard. For any undated references, its latest version (including all the modified documents) applies to this standard.

《Structural design code for pipelines of water supply and waste water engineering》 GB50332;

《Code for construction and acceptance of water and sewerage pipeline works》 GB50268;

《Code for design of gas transmission pipeline engineering》 GB50251;

《Code for construction and acceptance of city and town gas distribution works》 CJJ 33;

《Technical specification for detection and evaluation technique of urban sewer pipeline》 CJJ 181;

《Technical specification for trenchless rehabilitation and renewal of urban sewer pipeline》 CJJ/T 210;

《Technical specification for trenchless rehabilitation and replacement engineering of city gas pipe》 CJJ/T 147;

《Technical specification for trenchless rehabilitation and renewal of urban water supply pipelines》 CJJ/T 244;

《Technical specification for safety of operation, maintenance and rush-repair of city gas facilities》 CJJ 51;

《Technical specification for safety of urban sewer maintenance》 CJJ 6;

《Technical specification for operation, maintenance and safety of urban water supply pipe-networks》 CJJ 207.

© China Architecture & Building Press 2021
L. Wang et al., *Technology Standard of Pipe Rehabilitation*,
https://doi.org/10.1007/978-981-33-4984-1_2

Chapter 3
Basic Rules

3.1 Pipe Diameter Series

3.1.1 Small diameter: a pipe with inner diameter less than 200 mm.

3.1.2 Medium diameter: a pipe with an inner diameter greater than or equal to 200 mm and less than 800 mm.

3.1.3 Large diameter: a pipe with an inner diameter greater than or equal to 800 mm.

3.2 Pipeline Type

3.2.1 According to the type of pipeline, the pipeline can be divided into: drainage pipeline, water supply pipeline, gas pipeline, oil and gas pipeline, industrial pipeline, heat supply pipeline, etc.

3.2.2 The types of pipelines can be divided into: gravity pipelines and pressure pipelines according to the pipeline operating pressure status.

3.3 Pipeline Rehabilitation Length Series

Pipeline rehabilitation is divided into: localized rehabilitation and overall rehabilitation according to the length of one-time rehabilitation; overall rehabilitation is divided into: short distance, middle distance and long distance:

1. Localized/Spot: the rehabilitation length is less than 2 m;
2. Short distance: the rehabilitation length is less than or equal to 80 m;

© China Architecture & Building Press 2021
L. Wang et al., *Technology Standard of Pipe Rehabilitation*,
https://doi.org/10.1007/978-981-33-4984-1_3

3. Medium distance: the rehabilitation length is between 80 and 200 m;
4. Long distance: the rehabilitation length is more than 200 m.

3.4 Classification of Pipeline Rehabilitation Methods

3.4.1 Pipeline rehabilitation is divided into two categories: pipeline replacement and pipeline renovation.

3.4.2 According to the degree of pipeline structure rehabilitation, renovation is divided into: structural renovation, semi-structural renovation, and functional renovation.

3.4.3 Pipe replacement mainly includes: pipe cracking, pipe eating and pipe pulling.

3.4.4 Pipeline renovation are divided into two categories according to their integrity: localized renovation (spot renovation) and overall renovation.

3.4.5 According to the combination of the inner wall of the original pipeline and the outer wall of the liner, the pipeline renovation is divided into three categories: interval renovation, closed renovation, and pasted renovation.

3.4.6 According to whether the pipeline function or requirements (pressure, flow) are improved after the renovation, the pipeline renovation is divided into two types: upgrade renovation and non-upgrade renovation.

3.4.7 The overall renovation of the pipeline mainly includes: Sliplining (including Discrete Sliplining), pipe Segmental Lining, Improved Sliplining (including Swagelining, Folded Lining), Cured-in-Place-Pipe(CIPP) (including Inversion Lining and Drawing), Spirally Wound Lining, Spraying Lining, etc.

3.4.8 Localized renovation mainly includes: Spot CIPP, Stainless Steel foam sleeve, Spot grouting, pipe Segmental lining, etc.

3.5 Basic Principles of Pipeline Rehabilitation Method Selection

3.5.1 The construction purpose specified in the pipeline rehabilitation design requirements or assignment book (contract) should be met.

3.5.2 The pipeline rehabilitation should be determined according to the pipeline status, rehabilitation requirements and site environment requirements.

3.5.3 On the basis of adapting to the conditions and characteristics of the pipeline, priority is given to adopt more technologically advanced rehabilitation.

3.5.4 With the goal of high efficiency, low consumption, safety and environmental protection, ensure the quality of pipeline renewal, reduce labor intensity, and strive for good economic and social benefits.

3.5.5 The renovation should be adapted to the natural geographical and climate conditions of the construction area.

3.6 Selection of Pipeline Rehabilitation

3.6.1 Select the rehabilitation according to the diameter requirements after the pipeline rehabilitation. If you need to increase the diameter of the pipeline, you need to use the pipeline replacement, including the pipe creaking and the pipe eating, etc.

3.6.2 According to the type of pipeline renovation: structural renovation, semi-structural renovation, functional renovation, etc., select the pipeline renovation.

3.6.3 Select the pipeline rehabilitation according to the pipeline type. The appropriate pipeline types for different renovation methods are shown in Table 3.1.

3.6.4 Select the renovation method according to the old pipeline material. The suitable pipeline types for different rehabilitation methods are shown in the Table 3.2.

3.6.5 According to the pipe diameter and rehabilitation length, select the pipe rehabilitation method.

Table 3.1 Different types of pipelines are suitable for different renovation methods

	Method							
Type of pipeline	Sliplining	Pipe segmental lining	Modified sliplining	CIPP	Spirally wound lining	Spraying lining	Pipe cracking	Pipe eating
Sewer pipeline	++	++	++	++	+	++	++	++
Water supply pipeline	+	+	+	++	−	++	++	++
Gas pipeline	+	−	+	++	−	−	++	++
Oil and gas pipeline	+	−	+	++	−	+	++	++
Industrial pipeline	+	−	+	+	−	+	++	++

Note ++ very suitable; + suitable; − not suitable

Table 3.2 Rehabilitation methods for pipes of different materials

Type of pipeline	Method							
	Sliplining	Pipe segmental lining	Modified sliplining	CIPP	Spirally wound lining	Spraying lining	Pipe cracking	Pipe eating
Steel tube	++	++	++	++	+	++	++	++
Cast iron tube	++	++	++	++	−	++	++	++
Reinforced concrete pipe	++	++	++	+	−	++	++	+
Clay pipe	++	++	++	+	−	++	++	++
Plastic pipe	−	−	−	−	−	−	+	++

Note ++ very suitable; + suitable; − not suitable

3.6.6 According to the cross-sectional shape of the pipeline, select the pipeline rehabilitation method.

3.6.7 According to the degree of pipeline damage and the requirements after the rehabilitation, select the pipeline rehabilitation method.

3.6.8 Select the pipeline rehabilitation method based on the pipeline and its surrounding environment.

3.7 Main Process of Pipeline Rehabilitation

3.7.1 The construction process of the pipeline replacement project mainly includes:
Site investigation → Pipeline initial cleaning → Pipeline inspection → Building construction site → Work pit preparation → Equipment installation and commissioning → Replacement construction preparation → construction → Pipe end processing and connection → Testing and acceptance → Site environment restoration.

3.7.2 The construction process of the pipeline rehabilitation project mainly includes:
Site investigation → Pipeline initial cleaning → Pipeline inspection → Building construction site → Pipeline cleaning → Work pit preparation → Equipment installation and commissioning → Repair construction preparation → Rehabilitation construction → Pipe end processing and connection → Test and acceptance → Site environment restoration.

Chapter 4
Pipeline Rehabilitation Construction Organization Design

4.1 Basic Requirements for Construction Organization Design

1. Before implementing pipeline renovation, underground pipelines should be subjected to the necessary detection and inspection test to investigate their quality and operation, and the site data related to pipeline rehabilitation construction should be fully grasped;
2. Underground pipeline inspection should be carried out before the pipeline replacement project is constructed;
3. The construction organization design shall be carried out before construction of each project, and construction shall not be carried out without approval.

4.2 Design Basis for Construction Organization

1. Project pipeline rehabilitation design;
2. Pipeline quality inspection and evaluation report;
3. Technical requirements of pipeline rehabilitation project;
4. Relevant regulations, specifications and standards;
5. Contract or agreement.

4.3 Requirements and Main Contents of Construction Organization Design

4.3.1 Construction organization design requirements.

© China Architecture & Building Press 2021
L. Wang et al., *Technology Standard of Pipe Rehabilitation*,
https://doi.org/10.1007/978-981-33-4984-1_4

1. For major, technically complex or new technological pipeline rehabilitation projects, before implementation of the project design, test section should be carried out to evaluate the feasibility of the pipeline rehabilitation.
2. The optimization of the pipeline rehabilitation method should be based on the current national standards and specifications, based on full investigation and research, combined with the use of the pipeline and the opinions of the builder, to carry out technical comparisons of multiple schemes.
3. Before the design of the pipeline rehabilitation project construction organization, quality inspection and evaluation of the pipeline to be renewed should be carried out, and the pipeline site survey should be carried out at the same time to investigate and analyze the hydrogeological conditions and surrounding conditions of the pipeline in the construction area to fully grasp the site data related to the pipeline rehabilitation construction.
4. In the entire process of pipeline design and construction, the pipeline rehabilitation method and process are preferably selected to meet the requirements of the pipeline rehabilitation operation.
5. The pipeline rehabilitation construction organization design must meet the requirements of the management unit or construction unit and the contract agreement. Based on the site survey, it should be prepared based on the existing production, material consumption, personnel and equipment and cost quotas. According to the actual situation, equipment and optimal construction methods and processes can be selected to ensure project quality and obtain the best economic benefits.

4.3.2 Main contents of construction organization design.

1. Design standards and basis

2. Project overview

    (1) Explain the project name, work content, project deadline and construction requirements, etc.;
    (2) Explain the project's geographic location, traffic conditions, topography, local climate, etc.;
    (3) Explain the surrounding environment: surrounding structures and existing underground pipelines, site conditions, etc.;
    (4) If it is necessary to excavate the work pit, the engineering geological conditions should also be outlined.

3. Construction workloads

    (1) According to the pipeline to be repaired, explain the construction sequence;
    (2) Calculate the general workload according to the construction sequence.

4. Rehabilitation technical design

    (1) Determine the pipeline rehabilitation method according to the pipeline rehabilitation requirements;

(2) Determine the inspection and cleaning methods and standards according to the construction sequence;

(3) Determine the main construction materials for rehabilitation and main technical parameters;

(4) Design and calculation of the wall thickness of the liner;

(5) Perform the necessary calculation of the pipe ring stiffness strength;

(6) Give the main technical measures for pipeline rehabilitation;

(7) When new technology of rehabilitation is adopted, detailed technical measures shall be given;

5. Device selection

According to the diameter of the pipeline and the length of the rehabilitated pipeline, determine the type of related equipment, including cleaning equipment, testing equipment, power machine, pipeline rehabilitation construction related equipment, etc., the performance parameters of the main equipment should be noted.

6. Site layouts

(1) According to the site situation of the pipeline, give the layout of the work pit;

(2) According to the site investigation and the selected equipment, give the construction site layout.

7. Engineering quality requirements and guarantee measures

(1) Engineering quality index
According to the specific requirements of the pipeline rehabilitation design.

(2) Quality assurance measures

(1) To select the material of the inner liner, including protective measures and precautions during operation;

(2) According to the method of pipeline rehabilitation, technical measures to guarantee the construction quality are given;

(3) According to the specific conditions of the project, technical solutions for difficult point are given.

8. Accident prevention and safety technical measures

(1) According to the site conditions, the technical requirements and measures of cold prevention, fire prevention, flood prevention and other disasters and construction safety should be put forward;

(2) According to the pipeline situation, preventive and treatment measures for the rehabilitation quality accidents should be proposed;

(3) According to the existing pipelines around the work pit and underground, detailed protective measures for surrounding buildings and underground pipelines should be put forward;

(4) To propose emergency plans for possible major accidents.

9. Project construction arrangement

   (1) According to the above design, quantity and arrangement of construction equipment, labor, materials, etc. which are required for construction and emergency should be given;
   (2) Compile the time required for various projects and formulate the construction schedule;
   (3) Compile the construction schedule.

10. Construction organization and management

    (1) Give construction health and environmental protection measures;
    (2) For specific projects, determine the construction organization and formulate specific management measures.

4.4 Preparation and Approval of Construction Organization Design

4.4.1 Under the lead of the chief engineer of the construction unit, relevant technical personnel jointly prepare the rehabilitation construction organization design.

4.4.2 The construction organization design of pipeline rehabilitation shall be reviewed by the construction unit and reported to the task issuing unit or owner for approval.

4.4.3 During the construction of pipe rehabilitation, if the design is found to be inconsistent with the actual situation, promptly communication with the project issuing unit or owner should be done. Then modify the design in accordance with the relevant standards of project management, and do not alter the design without consent.

4.4.4 For pipeline rehabilitation projects with greater engineering difficulty and greater risks, industry expert demonstration meetings must be held to review and demonstrate the organization design of the pipeline rehabilitation project construction.

Chapter 5
Pipeline Cleaning

5.1 General Provisions

5.1.1 The cleaning involved in pipeline rehabilitation construction can be divided into cleaning for inspection and cleaning for construction.

5.1.2 The pipeline inspection and construction cleaning plan should be formulated based on the pipeline rehabilitation plan and the internal conditions of the pipeline.

5.1.3 The selection of pipeline cleaning method shall meet the requirements of corresponding pipeline cleaning quality.

5.1.4 After pipe cleaning, CCTV shall be used for cleaning quality inspection.

5.1.5 The cleaned pipeline shall be constructed in time or blocked and protected at both ends of the pipeline.

5.1.6 According to the accumulation of scale or sludge in the old pipeline, it can be divided into 5 levels according to the severity of the accumulation, see Table 5.1.

5.2 Pipeline Cleaning Quality Level

5.2.1 According to the requirements of pipeline renovation construction, the pipeline cleaning quality level can be divided into 3 levels, see Table 5.2.

5.2.2 In order to meet the quality requirements of pipeline cleaning, the construction unit can choose a combination of cleaning methods based on the inner conditions of the pipeline and the existing equipment process.

5.2.3 Pipeline renovation construction cleaning must meet the pipeline cleaning quality requirements for pipeline renovation construction.

© China Architecture & Building Press 2021
L. Wang et al., *Technology Standard of Pipe Rehabilitation*,
https://doi.org/10.1007/978-981-33-4984-1_5

Table 5.1 Classification of deposits inside old pipes

	Proportion of cross-sectional area taken by silt or dirt, R (%)
Level 1	R ≤ 10%
Level 2	10% < R ≤ 30%
Level 3	30% < R ≤ 50%
Level 4	50% < R ≤ 70%
Level 5	R ≥ 70%

Table 5.2 Pipeline cleaning quality grade classification table

Cleaning quality level	Applicable renovation method	Cleaning requirements	Supplementary requirements for steel pipe cleaning
Level 1	Pasted renovation (CIPP, spraying lining, etc.)	Dirt, obstacles, etc. in the pipeline are completely removed, the original material is exposed on the inner surface of the pipeline, there is no obvious protrusion, and the inner surface of the pipeline is kept clean	Reached national standard GB8923 "Steel surface modification level and rust removal level before painting" Sa 2.5
Level 2	Closed renovation (CIPP, swagelining, deformed lining, etc.)	Obstacles in the pipeline are completely removed, there are no sharp protrusions in the pipeline, the dirt has been basically removed, the residue should be firmly attached, and the surface of the residue should be smooth and the thickness should not exceed 2 mm	Reached national standard GB8923 "Steel surface modification level and rust removal level before painting" Sa 2
Level 3	Interval renovation (sliplining, pipe segmental lining, spirally wound lining, etc., including pipe cracking)	Obstacles in the pipeline are basically cleared, there are no sharp protrusions in the pipeline, and there are no attachments such as oxide scale, rust and paint coating that are not firmly attached	Reached national standard GB8923 "Steel surface modification level and rust removal level before painting" Sa 1

5.3 Pipeline Cleaning Methods

5.3.1 Manual cleaning/repairing.

1. Personnel enter the pipeline to clean or repair it with the aid of tools and tools, generally only used for cleaning or repairing the pipeline above DN800.
2. Manual cleaning/renovation process
Work area safety maintenance—pipeline preparation—ventilation and gas detection in the pipeline—manual cleaning—manual repair—cleaning and repair inspection—on-site recovery or preparation for follow-up work.
3. Manual cleaning/renovation should meet the following requirements:

   (1) Before entering the pipeline, the construction personnel should ventilate the pipeline in advance, detect the content of toxic and harmful gases and oxygen in the pipeline, and refer to the safety standard s for limited working space;
   (2) There should be personnel supervision on the ground during construction;
   (3) Upstream and downstream interception valves or other pipelines during construction should be completely closed and well-sealed. Communicate with the construction commander before the valve is opened before operation. When the drainage pipeline is sealed with airbags, someone should be sent to pay close attention downstream water level and airbag pressure.

4. Manual cleaning and repairing equipment use and maintenance

   (1) Equipment and tools for manual cleaning and repairing mainly include: grinder, exhaust fan, etc.
   (2) Check whether the electrical equipment such as grinders and cables are insulated and leakage, and whether ventilation equipment such as exhaust fans are working properly (explosion-proof electric tools), toxic gas detectors, flammable gas detectors, oxygen content detectors and other instruments are verified and working in good condition.

5.3.2 Robot clearing and removing obstacles.

1. It is controlled by personnel on the ground and can be automatically crawled and positioned in the pipeline. The robot with a special device completes special operations such as removing obstacles inside the pipeline, sharp protrusion grinding, and branch hole opening.
2. Robots can be used for special operations inside pipelines of various diameters and materials in short distances.
3. Robotic obstacle removal process: Safety maintenance in the operating area—pipeline preparation—pipeline pre-cleaning (if necessary)—robotic clearance and opening operations—post-operation inspection—robot recovery—site recovery or prepare for the follow-up work.

4. Use and maintenance of the robotic cleaning equipment: The main equipment of the robotic cleaning is the pipeline robot and perform regular maintenance on the robot according to the instructions of the robot manufacturer.

5.3.3 Cleaning with pigs.

1. Use the winch to drive the wire rope to pull or the hydraulic pressure to push the pipeline cleaning pig to scrape and bring out the dirt, sludge and other debris inside the pipe when walking in the pipe. This technology can clean long-distance pipelines. According to different types of dirt, hardness, quantity, etc. In the pipeline, choose the corresponding type of pipeline cleaning pig and the appropriate power winch or water and gas injection equipment. For cleaning in pipeline renovation projects, traction pig cleaning technology is mostly used.

2. Cleaning process of traction pig cleaning: winch in place, safe maintenance of working area—pipeline preparation—winch fixing and equipment commissioning, pipeline cleaning pig assembly—pipe threading—pipeline cleaning pig installation—repeated traction pig cleaning in the pipeline—inspection after cleaning—equipment recovery and evacuation—site restoration or preparation for follow-up work.

3. The flow cleaning of the pulling-type pig should meet the following requirements:

    (1) Choose the type and size of the pipeline cleaning pig according to the pipe material and the type of dirt. The pipeline cleaning pig must not damage the pipe to prevent the pig from being stuck in the pipe;

    (2) For pipelines with misaligned ports, if the misalignment is serious, the pipeline cleaning pig should be replaced or the pipeline cleaning pig installation should be stopped, and the pipeline cleaning pig should be installed after the misalignment is treated;

    (3) When cleaning the pipeline, the sewage and waste residue generated by the cleaning should be discharged from the inspection well or work pit;

    (4) Before the pipeline cleaning work, the reliability of the equipment should be carefully checked, including the tightness of the inflatable bladder and whether it is damaged, the drawability of the winch, and whether the wire rope is intact;

    (5) Ensure that there are no sharp debris, protrusions, etc. in the pipeline to prevent damage to the aerated or water-filled capsules;

    (6) The expansion pressure of the airbag in the pipeline should not be too large to prevent the pipeline from rupturing or deforming;

    (7) The advancing speed of the water bag or air bag in the pipeline should not exceed 0.1 m/s;

    (8) The debris cleaned from the pipeline shall be treated in accordance with the relevant standards (reduced and harmless treatment) and shall not be piled up or discarded at will.

4. Use and maintenance of the pulling-type pig equipment:

   (1) The pulling-type pig equipment mainly includes: traction winch and pipeline cleaning pig;

   (2) Before cleaning, carefully check whether the traction winch has sufficient traction, whether the wire rope is intact, whether the size of the pig is suitable, whether the assembly is correct, whether the traction winch is fixed firmly, and the two ends of the pipeline to be cleaned are unblocked;

   (3) Damaged steel wire ropes shall not be used continuously, and excessively thin steel wire ropes which do not match the traction force of the winch shall not be used and shall be replaced in time;

   (4) The pipeline cleaning pig of unsuitable type and size shall not be used and shall be replaced in time;

   (5) Carry out regular maintenance on the power part of the traction winch and the gearbox. After each cleaning, the steel wire rope should be promptly dehydrated and smeared with protective grease;

   (6) The pipeline cleaning pig should be cleaned and inspected, and damaged parts should be repaired and replaced in time.

5.3.4 High pressure water jet cleaning.

1. Use high-pressure water jet cleaning equipment to pressurize the cleaning water and form a high-speed water jet through the nozzle to crush and strip the dirt in the pipeline. The technology is mainly used for short-distance pipeline cleaning, and the corresponding pressure, flow rate and nozzle type are selected according to different types of dirt and hardness in the pipeline.

2. High-pressure water jet cleaning process: equipment in place, safe maintenance of the work area—pipeline preparation equipment connection and commissioning—repeated cleaning and dirt treatment of high—pressure water jets in the pipeline—inspection of equipment recovery and evacuation after cleaning—site restoration or preparation for follow-up work.

3. High-pressure water jet cleaning should meet the following requirements:

   (1) The water flow pressure must not cause damage to the pipe wall (such as erosion, grooves, cracks and perforations). When there are debris in the pipeline, the debris ejecting should be prevented in order to avoid the old pipe damage;

   (2) The jet water flow should not stay at a certain point of the pipeline inner wall for too long;

   (3) When cleaning the pipeline with high-pressure water jet, the sewage and waste residue generated by the cleaning should be discharged from the inspection well or work pit. In order to reduce the amount of water and environmental pollution, vacuum suction equipment and sewage purification recycling system should be used;

   (4) When the inner diameter of the old pipe is greater than 800 mm, manual entry into the pipe can also be used for high-pressure water jet cleaning;

(5) You can use high-pressure water to clean the inner wall of the pipeline, and use a fan and sponge ball to dry the wall after cleaning;

(6) The cleaned sewage and dirt should be collected and should be treated in a centralized manner;

(7) The inner wall of the pipe after cleaning should be dry and free from dirt. If it fails to meet the standard, it should be cleaned again.

4. Use and maintenance of high-pressure water jet cleaning equipment:

(1) The main equipment for high-pressure water jet cleaning is a high-pressure water jet cleaning vehicle;

(2) Before each use, you should carefully check whether the high-pressure hose is damaged, the joint connection is tight and sealed, and the damaged high-pressure hose cannot be used continuously, and should be replaced in time;

(3) High-pressure water jet cleaning operators should be trained before they can start operations;

(4) The high-pressure water jet cleaning vehicle high-pressure water pump set, high-pressure hose, nozzle and other accessories should be regularly checked to ensure that the equipment is complete and in good condition.

5.3.5 Sandblasting and derusting.

1. Using compressed air as the power to form a high-speed jet beam to spray the spray material (copper ore sand, quartz sand, emery sand, iron sand, Hainan sand) to the surface of the workpiece to be processed at high speed, so that the appearance or shape of the external surface of the workpiece changes. Due to the impact and cutting effect of the abrasive on the surface of the workpiece, the workpiece surface has a certain degree of cleanliness and different roughness, the mechanical properties of the workpiece surface are improved, the adhesion between them and the coating is increased, and the coating film is extended. The durability is also conducive to the leveling and decoration of paint.

2. Small diameter pipes below DN800 use rotary sandblasting for rust removal. Large diameter pipes—DN800 and above can be entered into the pipeline by construction personnel for manual sandblasting and rust removal.

3. Sandblasting and derusting process: air compressor, sandblasting tools, abrasives, etc. in place and safe maintenance of the operating area—pipeline preparation—pipeline pre-cleaning (if necessary)—sandblasting and derusting in the pipeline—dust removal—Inspection after derusting—on-site evacuation or preparation for follow-up work.

4. Sandblasting and derusting shall meet the following requirements:

(1) When the pipeline fouling is serious, the pipeline should be pre-cleaned, and after removing most of the dirt, sludge and obstacles in the pipeline, sandblasting and derusting should be carried out;

(2) The blasting abrasive used should meet the requirements;

(3) When manually entering the pipeline to remove sand and rust, the pipeline should be kept well ventilated;

(4) Recycle and reuse sandblasting abrasive as much as possible;

(5) Choose the type and size of the pipeline cleaning pig according to the material of the pipeline and the type of dirt. The pipeline cleaning pig must not damage the pipeline to prevent it from being stuck in the pipeline;

(6) Sandblasting should be carried out after mechanical cleaning is completed;

(7) Professional equipment (sandblasting system and sand suction system) should be used for cleaning. The sand suction equipment should have the ability to absorb 100% of the visible dust generated during the work;

(8) After sandblasting of the pipeline, the dust removal operation should be carried out. The inner wall of the pipeline should be dry, free of attachments, and 100% rust removed. The surface is roughened and exposed to metallic luster. It must reach the derusting level required by Table 5.2, If it fails to meet the standard, it should be cleaned up again;

(9) After the pipeline blast cleaning effect inspection is passed, the pipeline should be sealed before renovation.

5. Use and maintenance of sandblasting and derusting equipment:

(1) Sandblasting and derusting equipment mainly include: air compressor, sandblasting tank, sandblasting gun, sandblasting hose;

(2) Before sandblasting, carefully check whether the pressure and displacement of the air compressor are sufficient, whether the sandblasting tank is in good condition, and whether the pressure is sufficient; whether the sandblasting hose and gas delivery hose are intact; whether the sandblasting gun is suitable and effective;

(3) If manual blasting is used to remove rust, the equipment used for ventilation should also be checked for good operation.

5.3.6 Chemical cleaning.

1. Use chemical agents as cleaning agents, inject into the pipeline to circulate or soak. The chemical agents and the dirt in the pipeline undergo physical and chemical reactions to dissolve the dirt in the pipe. This technology is used for cleaning by physical methods that are difficult to remove or for pipes of special shapes.

2. Chemical cleaning process: injection pumps, chemical tanks and other equipment are in place, safe maintenance of the work area—pipeline preparation—cleaning process connection—chemical injection and cleaning—cleaning waste liquid discharge and treatment—washing after cleaning—Check after cleaning—Equipment recovery and evacuation—on-site to recover or prepare for follow-up work.

3. Chemical cleaning should meet the following requirements:

(1) Before chemical cleaning, ensure that the cleaned pipeline is tight and sealed;

(2) The appropriate type and formula of chemical cleaning agent should be selected to avoid corrosion of the pipeline body during the cleaning process;

(3)   The construction personnel should wear neat clothes to prevent chemical agents from harming the body;

(4)   Pay close attention to the spread of toxic and harmful gases generated during the chemical cleaning process to avoid personal injury;

(5)   After cleaning, the waste liquid should be recovered in time and handled in a proper manner, and no leakage or arbitrary discharge is allowed.

4.   Use and maintenance of chemical cleaning equipment:

(1)   The main equipment of chemical cleaning includes: pharmaceutical pump, medicine distribution tank, connecting pipe;

(2)   Before cleaning, carefully check whether the type, pressure, and displacement of the pharmaceutical pump are appropriate. After the cleaning system is connected, use fresh water to debug the system;

(3)   After cleaning, use fresh water to rinse the drug pump, dispensing tank, connecting pipe, etc. Regularly maintain the medicament pump set and the mixing tank.

5.4   Selection of Pipeline Cleaning Methods

5.4.1 Cleaning requirements.

1.   In order to check the quality of old pipelines, the cleaning quality must meet requirements of level 3 at least.

2.   When using the pasted renovation (CIPP, spraying lining, etc.), the pipe cleaning needs to meet the requirements of level 1.

3.   When the closed renovation (CIPP, swagelining, deformed lining, etc.) is adopted, the pipe cleaning needs to reach the requirement of level 2.

4.   When using the interval renovation (sliplining, spirally wound lining, and pipe segmental lining), the pipe cleaning needs to reach the requirements of level 3.

5.   When replacing with the pipe bursting and the pipe eating method, the pipe cleaning requirements should not affect the guide rod insertion and the broken tube head or the tube head moving forward.

6.   When spot renovation is used, the quality requirements of pipeline cleaning can refer to Table 5.2.

5.4.2 Choosing the pipe cleaning method is mainly according to the following factors: pipe types, pipe dirt types, pipe materials, general condition of pipe interiors.

Chapter 6
Pipeline Inspection and Quality Assessment

6.1 Pipeline Inspection

6.1.1 General provisions

1. Before inspection, the pipeline should be pre-processed as necessary, such as: interception, sewage suction, pumping, blowing and cleaning.
2. Inspection methods of pipeline quality include: manual inspection, camera inspection, periscope inspection, sonar inspection, magnetic flux leakage inspection, etc.

6.1.2 Pipeline inspection method

1. Manual inspection

    (1) Manual inspection is carried out by a person who enter the pipeline to visually detect, photograph, record and measure the internal conditions of the pipeline.
    (2) Manual inspection should meet the following requirements:
        (1) Manual inspection is only applicable to pipes with a diameter greater than 800 mm;
        (2) The manual inspection distance should not exceed 100 m each time;
        (3) When manual inspection is adopted, the depth of accumulated water in the pipeline shall not exceed 15% of the pipe diameter and shall not be greater than 0.5 m, and the flow velocity of the water in the pipe shall not exceed 0.3 m/s. or set measures such as temporary drainage to reduce the water level in the pipe;
        (4) The underground inspection staff should maintain communication with the ground staff;
        (5) The underground inspector should carry a camera to record and record the defect position in the pipeline in detail, and the picture should be clear.

© China Architecture & Building Press 2021
L. Wang et al., *Technology Standard of Pipe Rehabilitation*,
https://doi.org/10.1007/978-981-33-4984-1_6

2. Camera inspection

   (1) Camera inspection adopts Closed-Circuit-Television (CCTV) to inspect underground pipelines. The technicians interpret the pipeline status based on the inspection video.

   (2) The basic steps of inspection: collecting data—on-site survey—preparing testing plan—cleaning and blocking drainage—testing and collecting image data with CCTV testing system—summarizing data, outputting testing report—acceptance data accuracy—submission evaluation report.

   (3) Camera inspection should meet the following requirements:

       (1) Pipeline inspection and evaluation using closed-circuit television should be conducted in units of the pipeline between two adjacent inspection wells;

       (2) Before the inspection, the equipment should be thoroughly checked and debugged on the ground to ensure that the equipment can work normally;

       (3) Before the instrument enters the well for inspection, the display board should be photographed. The display board should display the information, such as the pipe location, pipe serial number, pipe diameter, inspection time and the name of the person in charge. The information displayed on the board should be written clearly in regular font.

       (4) When CCTV is used for inspection, the height of the water level in the pipeline should not exceed 20% of the vertical height of the pipeline;

       (5) When encountering defects or abnormalities in the pipeline, the inspection equipment should temporarily stop moving forward, change the camera to carefully photograph the abnormal parts of the defect, and then continue to move forward;

       (6) When inspecting obstacles that cannot be passed, the CCTV should be exited, and the obstacles should be cleared to continue the detection;

       (7) When scaling, siltation or severe corrosion and flaking on the inner wall of the old pipe affect the TV image effect, the inside of the pipe should be cleaned and continue to be tested;

       (8) The pipeline camera system should use color CCTV testing equipment with explosion-proof function and lens with rotation and zoom functions. The CCTV testing equipment should be equipped with a CCD camera with a horizontal resolution of not less than 460 lines;

       (9) CCTV inspection video data should be physically archived in a common format MPEG or AVI in a timely manner and be handed over to special personnel for collection and sorting.

3. Periscope inspection

   (1) Periscope inspection is also called Quick View inspection (QV). Quick View inspection (QV) is mainly used for rapid inspection and diagnosis of the

internal conditions of industrial containers or pipelines. During the inspection process, it can record and save the internal image of the inspected object in real time, which is suitable for pipelines with a diameter of 100–2000 mm.

(2) Pipeline periscope inspection steps: collecting data—on-site survey—preparing inspection plan—dredging and blocking drainage—using QV inspection system for inspection and collecting image data—summarizing data, issuing inspection report—acceptance data accuracy—submit an assessment report.

(3) QV inspection should meet the following requirements:

(1) QV inspection should be used for pipeline inspection and evaluation based on the pipeline section between two adjacent inspection wells;

(2) Before the test, the equipment should be thoroughly detected and debugged on the ground to ensure that the equipment can work normally;

(3) Before the instrument enters the well for inspection, the display board should be photographed, and the location, material, serial number, diameter, time, name of the person in charge, etc. of the pipeline to be detected should be clearly written on the display board with clear and correct fonts;

(4) When the QV is used for inspection, the water level in the pipeline cannot be full, and the periscope lens should be able to be put into the detection well to see the inside of the pipeline;

(5) Adjust the focal length of the QV inspection lens according to the distance, and gradually enlarge the focal length of the lens as the peeping distance increases;

(6) The QV inspection should adopt the explosion-proof function, the lens has the high-power zoom function, and the QV testing equipment should be equipped with a CCD camera with a horizontal resolution of not less than 460 lines;

(7) QV inspection video materials should be physically archived in a common format MPEG or AVI in time.

4. Sonar inspection

(1) Sonar inspection system is suitable for pipelines with high sewage fullness, large flow rate, and no drainage conditions. Traditional video inspection methods have not been able to achieve better inspection results. They are suitable for diameters (section size) from 125 to 6000 mm pipes of various materials within the scope. The pipeline profile sonar imager can accurately inspect a large number of structural defects (such as deformation, collapse, rupture, scaling, dark connection of branch pipes, etc.) and functional defects of pipelines (such as deposits and floating objects).

(2) Sonar inspection steps: collecting data—on-site survey—preparing the detection plan—threading—underwater scanning unit (sonar head) placed in the water in the pipeline—inspection by moving sonar head—analyzing

the data and issuing an inspection report—acceptance data accuracy—submit assessment report.

(3) Sonar inspection should meet the following requirements:

  (1) There must be water in the pipeline, sonar head can be completely submerged;
  (2) The pipeline cannot be completely blocked by siltation, and can be passed through ropes and sonar heads;
  (3) Make the sonar head move at a uniform speed according to the pipe diameter;
  (4) The sonar head should be kept fully immersed in the water during the entire inspection process;
  (5) When there is silt in the pipeline, in order to prevent the sonar head from being inserted into the mud, measures such as floating should be taken to ensure that the sonar head moves on the silt.

5. Magnetic flux leakage inspection

  (1) Magnetic flux leakage inspection technology can detect volume defects and pipeline characteristics caused by internal and external corrosion of the pipeline, and determine its size and accurate location. It has low requirements for the inspection environment and can be used for steel pipeline corrosion inspection such as oil pipeline, gas pipeline, and water pipeline. It is suitable for the long distance steel pipes with a diameter between 200–1000 mm.

  (2) Magnetic flux leakage inspection steps: collecting data—on-site survey—preparation of inspection plan—empty the pipeline before inspection—magnetic leakage inspection—analysis of data, preliminary inspection report—final inspection report—excavation verification data accuracy.

  (3) Magnetic flux leakage detection should meet the following requirements:

    (1) During inspection, the tube wall reaches full magnetic saturation;
    (2) Before inspection, the pipeline should be cleaned up to meet the requirement of installing an inspector;
    (3) Before the inspection, the pipeline undergoes geometric diameter measurement to ensure that the pipeline does not affect the large deformation, diameter reduction, and small curvature radius elbows of the detector;
    (4) When the detector is moving in the pipeline, it should keep a proper speed and move at a constant speed.

6.1.3 Pipeline rehabilitation inspection

1. Pipeline rehabilitation related inspection s include: preliminary inspection (preliminary judgment), detailed inspection (accurate judgment), pipeline cleaning quality inspection, engineering acceptance quality inspection, etc.

Table 6.1 Applicability of pipeline inspection methods

Detection method	Preliminary assessment and inspection	Detection before rehabilitation	Detection after rehabilitation
Manual inspection	Applicable	Applicable	Applicable
Camera inspection	Applicable	Applicable	Applicable
Sonar inspection	Applicable	Not applicable	Not applicable
QV inspection	Applicable	Not applicable/preliminary detection	Not applicable
Magnetic flux leakage inspection	Applicable	Applicable	Applicable

2. Pipeline quality inspection can adopt manual inspection, CCTV inspection, sonar inspection, QV inspection and other methods. The applicability of pipeline inspection methods is shown in Table 6.1.
3. The inspection unit shall collect relevant data in the area of the pipeline to be tested according to the requirements, organize technical personnel to carry out on-site surveys, master the site situation, formulate inspection plans, and make testing preparations.

4. The following information should be collected before pipeline inspection:

   (1) Technical data such as pipeline plans and completion drawings;
   (2) Existing historical data of the pipeline inspection;
   (3) Related pipeline data in the pipeline area to be inspected;
   (4) Engineering geological and hydrogeological data in the pipeline area to be inspected;
   (5) Other relevant information required for evaluation.

5. Before the pipeline inspection is conducted, site investigation should be carried out. The contents of the investigation are as follows:

   (1) Check the surrounding environmental conditions such as the geography, landform, underground pipelines, and traffic conditions in the pipeline area which is about to be tested;
   (2) Visualize the water level and flow, siltation and inspection of the structure of the well at the opening of the well;
   (3) Check the pipe position, inspection well position, pipeline depth, pipe diameter, pipe materials, etc. in the data;
   (4) Investigate the pipeline nodes and surrounding environment (above and below ground) according to the pipeline path;
   (5) Survey work pits, site layout and temporary bypass connections.

6. After the pipeline is cleaned, CCTV inspection of the pipeline should be conducted, and clear image data that can accurately locate and evaluate the pipeline defects should be saved record.
7. Before the pipeline rehabilitation construction, the basic requirements for pipeline inspection are as follows:

    (1) The pipeline to be inspected should be cleaned as necessary;
    (2) For gravity pipelines, it should be ensured that the water accumulated in the pipeline cannot exceed 15% of the pipe diameter.
    (3) For pressure pipelines, ensure that there is no water in the pipeline, and if there is a branch pipe, it should be blocked first;
    (4) Before the inspection starts, the pipeline must be cleaned, ventilated, and tested for toxic and harmful gases.

8. After the pipeline is repaired, CCTV inspection of the pipeline should be conducted, and clear image data that can accurately evaluate the quality of the pipeline should be saved record.
9. The basic requirements of pipeline inspection for pipeline completion acceptance are as follows:

    (1) The pipeline should be functionally tested, and the test records of the pipeline should be shown to the inspector;
    (2) For gravity pipelines, make sure that the water accumulated in the pipeline does not exceed 5% of the pipe diameter before inspection;
    (3) For pressure pipelines, make sure there is no water accumulation in the pipeline before inspection;
    (4) Before the inspection starts, the pipeline must be cleaned, ventilated, and tested for toxic and harmful gases.

6.2 Pipeline Quality Evaluation

6.2.1 General provisions

1. Pipeline quality evaluation is to evaluate the quality of the pipeline based on the pipeline quality inspection records and provide a reference for pipeline rehabilitation and maintenance decisions.
2. Pipeline quality evaluation takes the pipe section as the minimum evaluation unit, and an overall evaluation should be made when inspecting multiple pipe sections or areas.
3. The management unit or the owner should organize professional technical personnel or third-party professional detection agencies to evaluate the quality of the pipeline to determine whether the pipeline needs to be rehabilitated and its possible rehabilitation methods.

6.2.2 Gravity pipeline evaluation

1. For the evaluation of gravity pipelines, please refer to CJJ181 Technical Standard for Inspection and Evaluation of Urban Drainage Pipes.
2. Gravity pipeline evaluation is mainly based on pipeline embedding conditions, corrosion conditions, pipeline materials, and accident rates.
3. Gravity pipeline quality evaluation includes: functional defect and structural defect evaluation, defect type and defect grade evaluation, etc.
4. When the defect density index of the pipe functional defect or structural defect is greater than 0.3, an overall repair shall be carried out.
5. When the functional defect parameter is greater than 3, non-structural repairs should be performed.
6. When the structural defect parameter is greater than 3, semi-structural or structural repair should be performed.

6.2.3 Evaluation of pressure pipeline

1. The pressure pipeline quality evaluation for pipeline rehabilitation mainly carries out pipeline suitability evaluation. The pipeline suitability evaluation refers to the degree of influence of the defects in the pipeline on meeting the specified functional requirements, safety, and reliability, which can be dealt with according to the following 4 situations:

    (1) Defects that do not cause harm to production safety are allowed to exist;
    (2) No harm to safety, but the defects that will be further developed must be predicted for life and allowed to be used under monitoring;
    (3) If the defective component can be used for downgrade to ensure the safety and reliability requirements, it can be used for downgrade;
    (4) For defective components that pose a threat to safety and reliability, immediate measures should be taken to repair or stop using.

2. Applicability evaluation includes: pipeline failure analysis, pipeline residual strength evaluation, remaining life prediction, reliability analysis and risk management, etc. The applicability evaluation process is shown in Fig. 6.1.

3. Evaluation of pressure pipelines is mainly based on pipeline embedding conditions, corrosion conditions, pipeline materials, risk accident rates, and severity of accident consequences.

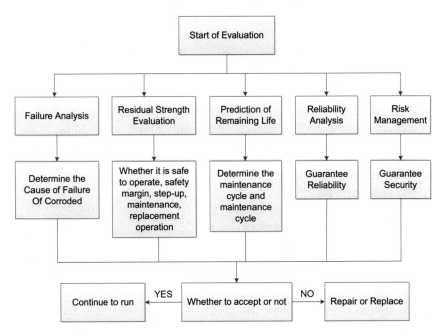

Fig. 6.1 The applicability evaluation process of pressure pipeline

Chapter 7
Pipeline Rehabilitation Design

7.1 General Provisions

7.1.1 The pipeline rehabilitation design should adopt the principle of design for different pipe section-by-section. For the pipe of different diameters and different defect levels, the pipeline rehabilitation should be designed section-by-section.

7.1.2 If it is confirmed by the external pressure failure load test that the rehabilitated pipeline has the same damage strength as the new pipeline, the pipeline renovation structure calculation can be omitted.

7.1.3 The design of the pipeline rehabilitation should meet the following basic requirements:

1. The rehabilitated pipeline should meet the requirements of the force;
2. The rehabilitated pipeline should meet the stability requirements;
3. The rehabilitated pipeline shall meet the requirements of over-current capability;
4. The rehabilitated pipeline shall meet the requirements of the dredging.

7.1.4 The basic principles of pipeline rehabilitation method selection are as follows:

1. Specify the type of pipeline rehabilitation: pipeline replacement, upgrade renovation, structural renovation, semi-structural renovation, functional renovation;
2. The main considerations include: pipeline type, pipeline quality, pipeline stress, flow area/diameter, pipeline shape, pipeline path (bending), surrounding environment (temperature, working pit, etc.), sanitation and resistance to the pipe For corrosion and other performance requirements, see Table 7.1.

7.1.5 In addition to the relevant requirements of this standard, the liner design in pipeline renovation should also meet the relevant standards or requirements of pipeline strength and stability in different industries.

© China Architecture & Building Press 2021
L. Wang et al., *Technology Standard of Pipe Rehabilitation*,
https://doi.org/10.1007/978-981-33-4984-1_7

Table 7.1 List of applicable conditions of pipeline rehabilitation methods

Method for renovation	Material of old pipeline	Shape of old pipeline	Maximum allowable bending of pipeline	Diameter of old pipeline (mm)
Pipe cracking	Not suitable for plastic tubes	Round	0	<600
Sliplining	No limit	No limit	11.25	No limit
Modified pipe sliplining	No limit	No limit	11.25	<1600
CIPP	No limit	No limit	45	No limit
Spraying lining	Not suitable for plastic tubes	No limit	22.5	No limit
Pipe segmental lining	No limit	No limit	No limit	>800
Spirally wound lining	No limit	No limit	0	No limit

Table 7.2 The new PE pipeline standard size ratio requirements for pipeline replacement

Covered soil thickness	SDR
0–5.0	≤21
>5.0	≤17

7.2 Pipe Replacement Design

7.2.1 When the old pipe is replaced by the pipe cracking method, the wall thickness of the pipe shall be designed in accordance with the requirements of the newly established pipes in different industries.

7.2.2 The maximum standard size ratio of the new PE pipeline replaced by the pipe cracking method should meet the requirements of Table 7.2.

7.3 Pipeline Renovation Design

7.3.1 See Table 7.3 for the check and calculation requirements of strength and radial stability of gravity pipeline renovation.

7.3.2 Refer to Table 7.4 for the check and calculation requirements of the strength and radial stability of the pressure pipeline renovation design.

7.3.3 For high-pressure and ultra-high-pressure pipelines, transport dangerous goods, and other dangerous pipelines, semi-structure renovation should not be adopted.

7.3.4 The methods, calculation formulas and parameter values for the design calculation and stability check of pipeline renovation strength shall be carried out with reference to the relevant provisions of engineering design codes of pipelines in different industries.

7.3.5 when the upgrading renovation is implemented, the upgraded pipeline technical requirements should be adopted, and the comprehensive strength of the new pipe or the composite with the old pipe and its grouting body or adhesive body of the annular gap should be carried out according to the requirements of the relevant pipeline standards in different industries check or radial stability check calculation to determine the liner thickness or grouting technical requirements.

Table 7.3 Basic requirements for check and calculation of gravity pipeline renovation strength and radial stability

Renovation type	Requirements and considerations of renovation	Renovation method	Checking method
Functional renovation	Renovation requirements: anti-corrosion, improving the performance of the inner surface of the pipeline, etc.; main considerations: the internal surface conditions of the original pipeline, the surface quality requirements after repair, the over-current capacity ratio before/after repair	Pasted renovation	No need for strength or stability check or verification calculation, determine the thickness of the liner according to requirements such as corrosion protection or surface treatment
		Closed renovation and interval renovation	Under the condition of vacuum negative pressure, check the strength or radial stability of the lined pipe to determine the thickness of the lined pipe
Semi-structural renovation	Renovation requirements: plugging and sealing, partial reinforcement of the structure, etc.; main considerations: ratio of over-current capacity before/after repair of the pipeline; external groundwater hydrostatic pressure or some external load required by the lining layer	Pasted renovation	Under common or partial external loads such as vacuum negative pressure and external groundwater static pressure, the strength check or radial stability check of the composite pipe of the new pipe and the old pipeline is performed to determine the thickness of the new pipe

(continued)

Table 7.3 (continued)

Renovation type	Requirements and considerations of renovation	Renovation method	Checking method
		Closed renovation and interval renovation	Under the action conditions of vacuum negative pressure and external ground water static pressure common or partial external load, the strength check or radial stability check of the composite pipe of the new pipe and the old pipe and its annular gap grouting body is performed to determine the new pipe thickness and grouting requirements
Structural renovation	Renovation requirements: overall structural repair; main considerations: ratio of overflow capacity before/after pipeline repair; external groundwater hydrostatic pressure, soil static load, surface vehicle live load, etc	Pasted renovation	Under the combined effect of external groundwater hydrostatic pressure, soil static load, surface vehicle live load, etc., the strength of the liner or radial stability check is performed to determine the liner thickness
		Closed renovation and interval renovation	Under the combined effect of external groundwater hydrostatic pressure, soil static load, surface vehicle live load, etc., the strength check or radial stability check of the composite pipe of the new pipe and the annular gap grouting body is performed to determine the new pipe thickness and grouting requirements

7.3.6 The value of vacuum negative pressure of gravity pipeline should be 0.03 MPa; the value of vacuum negative pressure of pressure pipeline should be 0.05 MPa.

7.3.7 The outer diameter of the inner liner used by sliplining should not be less than 10% of the original pipe inner diameter, and the reduction should not be greater than 50 mm.

7.3.8 The outer diameter of the inner liner of modified sliplining and CIPP should be consistent with the original pipe inner diameter.

7.3.9 When the liner pipe is located above the groundwater level, the standard size ratio (SDR) of the liner pipe in CIPP shall not be greater than 100; the standard size ratio (SDR) of the PE liner pipe shall not be greater than 42.

Table 7.4 Basic requirements for check and calculation of gravity pipeline renovation strength and radial stability

Renovation type	Requirements and considerations of renovation	Renovation method	Checking method
Functional renovation	Renovation requirements: anti-corrosion, improving the performance of the inner surface of the pipeline, etc.; main considerations: the internal surface conditions of the original pipeline, the surface quality requirements after renovation, the over-current capacity ratio before/after renovation	Pasted renovation	No need for strength or stability check or verification calculation, determine the thickness of the liner according to requirements such as corrosion protection or surface treatment
		Closed renovation and interval renovation	Under the condition of vacuum negative pressure, check the strength or radial stability of the lined pipe to determine the thickness of the lined pipe
Semi-structural renovation	Renovation requirements: plugging and sealing, partial reinforcement of the structure, etc.; main considerations: ratio of over-current capacity before/after renovation of the pipeline; internal pressure required by the lining, external hydrostatic pressure of the groundwater, or some external loads, etc	Pasted renovation	Under the combined conditions of internal pressure or vacuum negative pressure and external groundwater static pressure or part of external load, the strength check or radial stability check of the composite pipe of the new pipe and the old pipeline is performed to determine the thickness of the new pipe

(continued)

Table 7.4 (continued)

Renovation type	Requirements and considerations of renovation	Renovation method	Checking method
		Closed Renovation and interval renovation	Under common or partial external loads such as internal pressure or vacuum negative pressure and external groundwater static pressure, perform the strength check or radial stability check of the composite pipe of the new pipe and the old pipe and its annular gap grouting body to determine the thickness of the new pipe and the grouting requirements
Structural renovation	Renovation requirements: overall structural repair; main considerations: ratio of overflow capacity before/after pipeline renovation; internal pressure, external groundwater hydrostatic pressure, soil static load, surface vehicle live load, etc	Pasted renovation	Under the combined effect of internal pressure or external groundwater hydrostatic pressure, soil static load, and surface vehicle live load, etc., perform the strength check or radial stability check of the liner to determine the thickness of the liner
		Closed renovation and interval renovation	Under the combined effect of internal pressure or external groundwater hydrostatic pressure, soil static load, and surface vehicle live load, etc., perform the strength check or radial stability check calculation of the composite pipe of the new pipe and the grouting body with annular gap to determine the thickness of the new pipe and the grouting requirements

7.3.10 When the interval renovation is used for semi-structural and structural renovation, the annular gap between the liner pipe and the old pipe should be grouted. The liner pipe, grouting body and the composite pipe body formed with the old pipe should be checked and tested whether can bear the total load acting on the pipe.

7.4 Calculation of the Maximum Tensile Force of the Inner Liner

7.4.1 When the modified sliplining is used to renovate the pipeline, the maximum allowable pulling force of the liner should be calculated according to (7.1):

F_{max}—Maximum allowable pulling force of inner liner;

D_0—Standard outer diameter of inner liner (mm);

D_i—Standard inner diameter of inner liner (mm);

K—Safety factor, take 3;

σ—Liner pipe tensile yield strength, PE80 pipeline is 17.0 N/mm², PE100 pipeline is 21.0 N/mm²

$$F_{max} = \frac{1}{K} \cdot \frac{1}{4}\pi\sigma(D_0^2 - D_i^2) \tag{7.1}$$

7.4.2 When the pulling process is used for sliplining, the maximum allowable pulling force of the liner should be calculated according to (7.1), K is 2.

Chapter 8
Preparations Before Construction

8.1 General Provisions

8.1.1 Before we design the engineering of the trenchless pipeline rehablitation, relevant information on the construction site should be collected and site surveys should be conducted.

8.1.2 For the pipeline rehabilitation project under special conditions, the feasibility of the process method can be verified using the test section engineering.

8.1.3 The validity period of the pipeline rehabilitation design of trenchless shall not exceed one year, and projects over one year shall be reviewed or redesigned.

8.1.4 Subject to the conditions of construction, the construction site shall occupy no or less cultivated land, forest land, green land and site as much as possible.

8.1.5 The distribution of underground cables, pipelines and high-voltage wires on the construction site should be understood.

8.1.6 Formulate reasonable platooning schemes, protection schemes for underground facilities and traffic dredging schemes, etc.

8.2 Site Investigation

8.2.1 The depth and diameter of all inspection wells on the old pipeline, as well as the branch pipes in the well, the flow direction, the elevation of the bottom of the pipe, the diameter of the inlet and outlet, etc. should be investigated.

8.2.2 Detailed investigation shall be conducted on the conditions of other underground pipelines in the construction area, soil conditions in the construction area, underground wells in the construction area, civil air defense facilities, obstacles,

© China Architecture & Building Press 2021
L. Wang et al., *Technology Standard of Pipe Rehabilitation*,
https://doi.org/10.1007/978-981-33-4984-1_8

above-ground buildings, transformers, telephone poles and green spaces in the construction area.

8.2.3 The location, trend, inner diameter, length, material, buried depth, interface type of the pipeline to be repaired, the number and location of valves, fittings and branch pipes on the pipeline, the bending situation of the pipeline, the working pressure of the pipeline, the previous construction technology, maintenance, operation status, construction time, etc. The inner diameter or section size, perimeter, ellipticity, and pipeline length should be remeasured.

8.2.4 The bending condition and radius of curvature of the old pipe should be investigated.

8.2.5 The pipeline defects to be repaired (leakage, corrosion, holes, cracks, etc.) should be investigated or reviewed and an accurate assessment of the pipeline structure and function should be made.

8.2.6 The investigation should be clearly determined whether the renovation of the water supply pipeline must be partially renovated and pretreated.

8.2.7 The entry and exit road survey aims to know about the road traffic in different weather and different time periods, and find out the impact of the flow of vehicles and pedestrians on the entry road and the construction site.

8.2.8 Water use survey should contain investigatation about the distance between the water source of the existing water source and the construction site, the amount of water, the laying method of the transmission line, etc.

8.2.9 The electricity consumption survey should contain the investigatation of the power, voltage, location of the power supply and the distance between the construction site and the transmission line laying method.

8.3   Underground Pipeline Detection

8.3.1 When pipeline replacement is used for pipeline rehabilitation, it is necessary to carry out detection on surrounding underground pipelines.

8.3.2 The objects of underground pipeline detection shall include various pipelines buried underground, such as water supply, drainage, gas, heat, industry, etc., as well as electric power and telecommunication cables.

8.3.3 The size, direction and burial depth of other underground pipelines in the construction area shall be verified. When the construction pipe section is close to other pipelines and facilities (such as: natural gas, power cables, communication optical cables, etc.), you should contact the relevant units to take safety measures.

8.3.4 Before on-site detection of underground pipelines, the existing underground pipeline data and related surveying and mapping data in the survey area should be fully collected and sorted, and should include the following contents:

(1) Various existing underground pipeline diagrams;
(2) Design drawings, construction drawings, completion drawings and technical description materials of various pipelines;
(3) Topographic map of corresponding scale;
(4) The coordinates and elevation of the measurement area and its adjacent measurement control points.

8.4 Construction Site Preparation

8.4.1 The construction site should ensure road access and the site is level.

8.4.2 The construction site should meet the requirements of different construction equipment layouts on the site area.

8.4.3 Traffic guidance investigation should be conducted for road construction, and special plan for road construction should be approved.

8.4.4 Site survey should includes the investigation of the construction equipment and vehicle parking location.

8.4.5 The location of the work pit shall be determined according to the design scheme and the actual situation on the site. The location of the work pit shall avoid underground structures, underground pipelines and other obstacles.

8.4.6 The slope of the working pit or working well shall be supported before construction. When the construction in the rainy season, the pipeline is buried deep, and the excavation location is on the municipal road, in order to ensure the construction safety, the support method of bolt shotcrete should be adopted.

8.4.7 When there is dynamic load around the work pit, corresponding protective measures should be taken according to the environmental conditions. During the rehabilitation construction process, if it is necessary to pile earthwork, materials, construction and placement of construction machinery and equipment near the work pit, the safety of the work pit support should be checked and confirmed to be safe.

8.4.8 Prior to the construction of the gas pipeline renovation, a plan should be formulated for the gas suspension, evacuation, and purge of the pipeline in service, and it should comply with the provisions of the urban gas-related safety technical standard s.

8.4.9 Pipeline pretreatment includes: local reinforcement, repair, plugging, etc. Pipeline cracks, interface dislocations, grooves, sharp protrusions, etc. are repaired in accordance with the design and construction process requirements. If necessary, point-shaped excavation can be used to remove obstacles that affect the construction of the new pipe.

8.4.10 Welding of plastic pipes: welding procedure evaluation shall be carried out before welding, and the welding quality shall comply with the relevant regulations of relevant industries.

Chapter 9
Pipe Cracking

9.1 General Provisions

The pipe cracking is suitable for the replacement of concrete pipes, steel pipes, and cast iron pipes whose original diameter is DN100 ~ 600 mm, and suitable for the replacement of diameter expansion construction from first grade to second grade.

9.2 Construction Design and Materials

9.2.1 When the pipe cracking is adopted, the top soil of the old pipe to be rehabilitated shall not be less than 0.7 m. The safety distance between the outer wall of the new pipeline to be laid and the outer wall of other underground pipelines shall meet the following requirements:

1. The safety distance of the pipes with the same diameter should not be less than 0.5 m;
2. The safety distance of the expansion pipes and the other pipes shall not be less than 0.7 m.

9.2.2. For new pipes, HDPE new pipes of PE80 grade or above should be generally selected, and they should comply with the relevant material standards.

9.2.3 According to different use function, pressure level, pipeline buried depth, medium, etc., HDPE pipes with different standard size ratio SDR values should be determined.

9.2.4 In the same rehabilitated pipe section, the same HDPE pipe shall be used, and pipes of different models or different manufacturers shall not be used.

9.2.5 New pipes should be carefully inspected, and there should be no visible cracks, holes, scratches, inclusions, bubbles, deformation or other defects.

© China Architecture & Building Press 2021
L. Wang et al., *Technology Standard of Pipe Rehabilitation*,
https://doi.org/10.1007/978-981-33-4984-1_9

9.3   Equipment Use and Maintenance

9.3.1 The specifications of the pipe cracking equipment and its applicable pipe diameter should be selected according to Table 9.1.

9.3.2. The recommended split pipe head (knife) for different pipes should be selected according to Table 9.2.

Table 9.1 List of the recommended pipe cracking equipment

Pulling force of pipe cracking equipment (kN)	Work pit / inspection well	Applicable pipe diameter range (mm)
200	√	50 ~ 150
400	√	50 ~ 300
770	√	65 ~ 450
1250	√	150 ~ 600
2500	√	300 ~ 600

Note "√" means applicable

Table 9.2 List of the recommended split pipe head (knife)

Power mode	Hydrostatic				Pneumatic
Rolling cutter Type / Pipe type	Rolling cutter	Lead pipe cutter	Plastic pipe cutter	Cutter with expanding head	Cutter with expanding head
CIP(Cast Iron Pipe)	√	–	–	√	√
DIP(Ductile Iron Pipe)	√	–	–	–	–
Steel Pipe	√	–	–	√	√
PE/PP Pipe	–	–	√	√	√
PVC Pipe	–	–	–	√	√
ACP(Asbestos Cement Pipe)	–	–	–	√	√
Lead Pipe	–	√	–	–	–
VCP(Vitrified Clay Pipe)	–	–	–	√	√
CP(Concrete Pipe)/RCP(Reinforced Concrete Pipe)	√	–	–	√	√
GRP(Glass Fiber Reinforced Polyester Pipe)	√	–	–	√	√
Brick Pipe	–	–	–	√	√

Note "√" means applicable; "–" means not applicable

9.3.3 Equipment maintenance.

1. Wipe the guide rod, quick-install rod, expansion device and pipe expansion equipment, etc., and apply anti-rust agent such as lubricating oil, and check the surface of the parts for cracks and other damage.
2. Check the appearance, wear, and damage of the splitting head (knife), record the wear data, replace and repair the worn parts in time.

9.4 Construction Preparation

9.4.1 The pit position and size of the work pit shall meet the following requirements:

1. The size of the work pit should be selected according to the size of the equipment to facilitate the construction operation. The bottom of the pit should be 350 mm lower than the bottom end of the outer wall of the old pipeline to be rehabilitated;
2. The initial working pit width of the inlet pipe should be greater than the diameter of the new pipe 300 mm and not less than 650 mm; the length should be determined according to the allowable bending radius of the PE pipe, see Fig. 9.1.

9.4.2 The construction of the work pit shall meet the following requirements:

1. The wall surface of the equipment pit should be perpendicular to the plane of the bottom of the pit and the center line of the old pipe;
2. Gravel not less than 80 mm thick should be laid under the equipment pit;
3. The interface between the inlet pit wall and the old pipeline should be properly excavated;
4. The work pit should be kept dry, and precipitation should be carried out if necessary.

9.4.3 When using hydrostatic method for pipeline rehabilitation construction, the following requirements shall be met:

1. The alarm device and emergency switch of the equipment should be tested before construction;

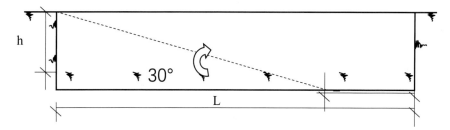

Fig. 9.1 Schematic diagram of the initial working pit of the inlet pipe. h—Working pit height; L—working pit length

Fig. 9.2 Schematic diagram
of circular cutter position

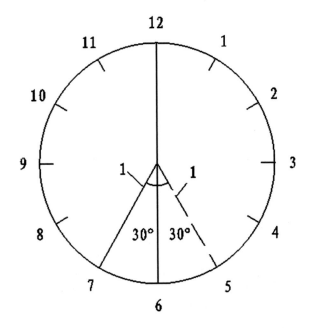

2. Check the safety and reliability of the equipment;
3. The pulling rod and the new pipeline should be firmly connected;
4. The connecting end of the pipe splitter should be intact and connected to the pulling rod;
5. When the pipe splitter enters the old pipeline, the position of the cutter wheel should be at an angle of 30° with the diameter of the pipeline perpendicular to the ground, see Fig. 9.2.

9.4.4 Before welding, the polyethylene pipe should be evaluated for welding procedures to ensure the welding quality.

9.5   Pipe Cracking Construction

9.5.1 The staff in the starting work pit should be coordinated with the operation host personnel at the end of the equipment pit. Intercom communication should be adopted.

9.5.2 When using the pneumatic method for pipeline rehabilitation construction, the following requirements shall be met:

1. During the process of cracking the pipe, a constant pulling force should be applied to the cracked pipe head;

2. Before the cracked pipe head reaches the receiving pit, the construction should not be terminated.

9.5.3 During construction, the new pipe should be placed on the roller bracket on the ground, and it is strictly prohibited to drag on the ground.

9.5.4 When there is a sudden increase in traction during construction, the construction must be stopped immediately and the cause must be ascertained before proceeding.

9.5.5 During construction, the number of the pulling rods entering the pipeline shall be recorded to determine the position of the pipe splitter when it moves forward.

9.5.6 After the new pipe is pulling into the old pipe, the excess pipe section should be cut off, and 500 mm length pipe sections should be reserved at both ends to adapt to the later contraction of the pipe.

9.5.7 The polyethylene pipes in the working pit shall relieve the stress for not less than 24 h before connection.

9.5.8 In the starting pit and receiving pit, the annular gap between the new pipeline and the soil shall be sealed, and the sealing length shall not be less than 200 mm.

9.5.9 After the polyethylene pipes in the work pit are connected, the leak detection shall be carried out, and the work pit shall be backfilled after passing the leak detection.

9.6   Quality Assurance Measures

9.6.1 Equipment operators should strictly abide by the equipment operation rules.

9.6.2 All pipes must comply with the relevant standards of various industries. Before the material is used, the material performance should be re-inspected according to relevant regulations.

Chapter 10
Sliplining

10.1 General Provisions

10.1.1 It is suitable for the renovation of various specifications of round, square and other special-shaped pipes.

10.1.2 When partial damage occurs on the pipe, such as falling off and missing partially, misalignment of pipe end, partially cracked, partially corroded, partially leaked, etc., localized repair or pretreatment should be performed first.

10.1.3 Each renovated pipe section should be decided whether it needs to be designed separately according to the actual situation. If the bend of a certain section of the pipe section exceeds the capacity of the repaired bent pipe or the pipe diameter changes, section measures should be taken. Those who need to design work pits can set up work pits at these points as the receiving pit of the previous section of pipeline and the starting pit of the next section of pipeline.

10.2 Construction Design and Materials

10.2.1 The outer diameter of the liner used in sliplining should be smaller than the inner diameter of the original pipe, but the reduction in diameter should not exceed 10% or 50 mm of the inner diameter of the old pipe.

10.2.2 The minimum length of the initial working pit of the pulling process shall be calculated according to formula (10.1), and the layout of the initial working pit shall be as shown in Fig. 10.1.

$$L = [H \times (4R - H)]^{\frac{1}{2}} \tag{10.1}$$

In the formula:

© China Architecture & Building Press 2021
L. Wang et al., *Technology Standard of Pipe Rehabilitation*,
https://doi.org/10.1007/978-981-33-4984-1_10

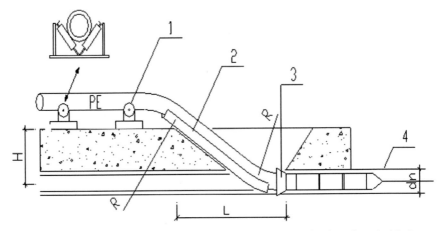

1 ground roller frame; *2* wear pad; *3* trumpet-shaped introduction pipe; *4* old pipe

Fig. 10.1 Schematic diagram of the initial working pit layout of the traction process

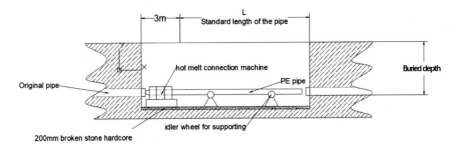

Fig. 10.2 Schematic diagram of the initial working pit layout of the jacking process.

L length of working pit (m);
H Buried depth of pipeline (m);
R Allowable bending radius of polyethylene pipe (m), and $R \geq 25d_n$.
d_n Outer diameter of new tube (m).

10.2.3 The initial working pit size and layout of the jacking process are shown in Fig. 10.2.

10.3   Equipment Use and Maintenance

10.3.1 The equipment used includes: traction device, hot melt welding machine and grouting equipment. The traction device is generally composed of a winch, a guide pulley, a wire rope, a traction head, etc. The traction system should be equipped with an automatic display device to record the traction force during construction.

10.3.2 The equipment required on site includes conventional equipment and special equipment. Conventional equipment is used and maintained in accordance with relevant regulations and instructions. Special equipment must be used and maintained in accordance with the equipment instruction manual.

10.4 Construction Preparation

10.4.1 It should be selected according to the design plan and the actual situation on site, develop a try-on plan, and determine the location of the work pit.

10.4.2 The cleaning and testing of pipelines before pipe penetration shall comply with the provisions of this code.

10.4.3 Before passing through the pipe, a polyethylene section with a length of not less than 3 m and the same diameter as the liner pipe shall be used to detect the passing capacity of the pipe section, and the depth of scratches on the surface shall be tested. The depth of the scratches shall not exceed 10% of the wall thickness.

10.4.4 The connection of the pipeline shall meet the following requirements:

1. Check the damage of the pipeline before connection: the score on the outer surface of the pipeline should not exceed 10% of the wall thickness, there should be no buckling caused by excessive or sudden bending, the flattening rate for short pipes should not exceed 5%, and the inner surface should not have any wear and cutting;
2. For the use of open flame connection equipment to connect pipes in work pits or inspection wells, the content of combustible gas should be evaluated in advance;
3. The connection of pipelines should adopt the method of electro fusion welding butt joint. The electro fusion welding butt joint shall meet the requirements in GB19809 of "Code of Practice for the Preparation of Plastic Pipes and Fittings Polyethylene (PE) Pipes / Tubes or Pipes / Pipes electro fusion welding Butt Components"

10.5 Sliplining Construction

10.5.1 The new pipe can be installed into the original pipe by pulling, pushing or combining pulling and pushing. In the installation of ultra-normal wall thickness or ultra-long polyethylene pipes, the combination of pulling and pulling should be used.

10.5.2 The pulling process should meet the following requirements:

1. When pulling the pipeline, necessary measures should be taken to prevent the liner from being crushed and scratched when entering the old pipeline;
2. In the case of pulling obstruction, the size of the drawing force should be controlled to prevent the pipeline from exceeding the rated allowable value;
3. The pulling rate of the pipeline shall not exceed 1.5%, the pulling speed shall not exceed 0.3 m/s, and the construction shall be slowed down in the pipeline where the bend or the pipeline deformation is large;
4. The pulling operation process should not be interrupted;
5. When inserting the new pipe into the old pipe, the length of the end of each section of the new pipe protruding from the port of the old pipe should meet the requirements of the pipeline's tensile deformation recovery and connection operation. The recommended value is: 1% L + 20 mm; The recovery time should be maintained for 24 h.

10.5.3 The short pipe pushing process should meet the following requirements:

1. Electro fusion welding welded polyethylene linning pipe can also be installed by pushing;
2. When the space of the working pit is restricted or a mechanical socket type joint is used, the pushing process can be used to allow water flow in the original pipeline, but its water level should be below the pipeline arch line;
3. This process can also be used for the curve jacking of short tubes;
4. Water pumps and temporary drainage facilities should be prepared at the construction site.

10.5.4 Reinforcement measures such as fixing and grouting can only be carried out after the new pipe is stable and the stress is completely relaxed.

10.5.5 The annular gap between the liner pipe and the original pipeline should be grouted. The grouting should meet the following requirements:

1. For a pipe with a diameter of more than 800 mm, the annular gap between the new pipe and the old pipe must be grouted;
2. Grouting should be carried out in sections, and the annular gap should be evenly filled, and the grouting pressure should not be greater than the allowable grouting pressure of the new pipe;
3. During the grouting process, the grouting pressure should be adjusted by installing vertical pipes or other methods;
4. When the diameter of the new pipe is greater than 900 mm, it shall be supported in the pipe to prevent the new pipe from deforming under the effect of grouting pressure.
5. After the grouting is completed, the grouting hole should be sealed.

10.6   Quality Assurance Measures

10.6.1 The materials used should conform to the relevant national codes and standards and be re-inspected according to the codes.

10.6.2 Sliplining construction shall be functionally tested in accordance with relevant technical regulations and acceptance specifications.

Chapter 11
Pipe Segmental Lining

11.1 General Provisions

11.1.1 The pipe segmental lining is divided into modular lining and stainless-steel lining; the modular lining is only applicable to the renovation lining gravity flow pipes. The material for the module lining is PVC sheet profile, and the material for the stainless-steel lining is stainless-steel plate.

11.1.2 The pipe segmental lining is only applicable to the repair of pipelines that can be accessed by people and have a caliber greater than 800 mm.

11.1.3 The pipe segmental lining can be applied to the repair of the following situations, see Table 11.1.

11.2 Construction Design and Materials

11.2.1 The material used in the pipe segmental lining is a factory-prepared stainless steel, PVC and other sheet profiles with durability, corrosion resistance and smooth surface.

11.2.2 Module lining.

1. Commonly used modular lining PVC sheet profile structure is shown in Fig. 11.1;
2. Refer to Table 11.2 for the recommended pipe segment size;
3. Refer to Table 11.3 for the recommended size of rectangular pipe segments;
4. The sheet profiles used in the same repaired section should be the same material, and there should be no defects such as visible cracks, holes, entrained debris or other damage.
5. Samples of different profiles of different production batches should be sampled and tested separately. The detection and requirements of the physical properties of PVC sheet profiles shall meet the requirements of Table 11.4.

© China Architecture & Building Press 2021
L. Wang et al., *Technology Standard of Pipe Rehabilitation*,
https://doi.org/10.1007/978-981-33-4984-1_11

Table 11.1 List of applicable scope of pipe segmental lining

Item	Scope of application
Repairable object	Round, rectangular, horseshoe-shaped
Repairable size	Round pipe: diameter 800 ~ 3000
	Rectangular pipe: 1000 × 1000 ~ 2200 × 2200
Construction length	Unlimited
Construction water environment	Module lining: water depth less than 500 mm; stainless steel lining: dry without water
Pipeline interface longitudinal misalignment	Less than 2% of the diameter
Pipeline interface is laterally dislocated	Under 150 mm
Radius of curvature	Above 8 m
Pipeline bend	Under 3°
Tilt adjustment	Adjustable height below 2% of diameter
Working surface	Over 30 m² during assembly; over 35 m² during grouting

Fig. 11.1 Schematic diagram of the sheet profile used for the pipe segmental lining

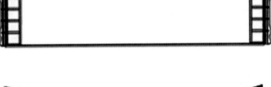

6. The pipe segment profile for modular lining shall be clearly marked, and the mark shall include the manufacturer's name or trademark, product number, place of origin, production equipment, production date, model, material grade, and the name of the specification on which the product is produced. Each piece of segments profile should be marked.

7. The adhesive or sealant used should be compatible with the PVC composite material and pipe assembly process.

8. The strength and fluidity of the grout used in the module lining must meet the requirements of Table 11.5 of this regulation. At the same time, the grout must also have anti-segregation, micro-expansion, anti-cracking and other properties.

9. The filler is not easy to separate in water and has excellent fluidity and strength. The formulation and composition of the filler are shown in Table 11.6.

11.2.3 Stainless Steel Lining.

1. The stainless steel plate should comply with the relevant provisions of the current national standard "Stainless Steel Cold Rolled Steel Plates and Strips" GB /

Table 11.2 Recommended size list for round pipe segments

Original pipe diameter (mm)	Module			
	Number of divisions	Diameter after renovation (mm)	Height of Module (mm)	Thickness of Module Surface (mm)
800	4	725	23.5	4.5
900	4	820	24	4.5
1000	4	915	239	5
1100	4	1005	29	6
1200	4	1105	29	6
1350	4	1240	32	6
1500	5	1370	36	6
1650	6	1510	39	6
1800	6	1650	43	6
2000	8	1840	48	6
2200	8	2030	53	6
2400	8	2220	58	6
2600	9	2405	60	6
2800	10	2590	68	6
3000	10	2775	73	6

Note The width of all the module is 200 mm. Only for the module whose diameter is 1500 mm, its width is 186 mm

Table 11.3 List of recommended dimensions for rectangular pipe segments

Original pipe diameter (mm)	Module			
	Number of divisions	Diameter after renovation (mm)	Height of Module (mm)	Thickness of Module Surface (mm)
1000 X 1000	12	895 X 895	40	6
1100 X 1100	12	986 X 986	40	6
1200 X 1200	8	1076 X 1076	40	6
1350 X 1350	12	1225 X 1225	40	6
1500 X 1500	12	1375 X 1375	40	6
1650 X 1650	16	1525 X 1525	40	6
1800 X 1800	16	1675 X 1675	40	6
2000 X 2000	16	1875 X 1875	40	6
2200 X 2200	16	2075 X 2075	40	6

Note The module width is 200 mm

Table 11.4 Physical requirements of PVC pipe segment materials

Material performance	Test Methods	Minimum value
Density	/	$\geq$1.6 g/m^2
Longitudinal tensile strength	GB/T 1040.2	$\geq$44.4 MPa
Longitudinal bending strength	GB/T 9341	$\geq$75 MPa
Thermoplastic vicat softening temperature	GB/T 1633	$\geq$75.4 °C
Thermoplastic pipe material withstands external impact (TIR)	GB/T 14,152	0%

Table 11.5 Basic requirements of grouting

Structural performance	Minimum value
Compressive strength	$\geq$30 MPa
Mobility	$\geq$270 mm

Table 11.6 Formulation and composition of Filling Material

Material	Composition	Weight Ratio
Cement	Blast Furnace Cement	1722
Sand	Limestone crushed stone with a maximum particle size of 1.2 mm	
Mixture	Low shrinkage material + water reducer + defoamer + tackifier	
Water		365

T3280 and the national standard "Stainless Steel Hot Rolled Steel Plates and Strips" GBT4237, and the performance of the welding consumables should comply with the current national standards "Stainless Steel Electrodes" GB / T983 relevant regulations.

2. The mechanical performance of stainless-steel plates liner with different brands should be in accordance with Table 11.7.

Table 11.7 Mechanical performance of stainless-steel plates

Grade	Performance		Test reference standard
06Cr19Ni10 (type 304)	Pipe tensile strength	$\geq$520 MPa	"Tensile test of metal materials Part 1: Test method at room temperature" GB / T228.1
	Pipe elongation	$\geq$35%	
022Cr19Ni10 (type 304L)	Pipe tensile strength	$\geq$480 MPa	
	Pipe elongation	$\geq$35%	
06Cr17Ni12Mo2 (type 316)	Pipe tensile strength	$\geq$520 MPa	
	Pipe elongation	$\geq$35%	
022Cr17Ni12Mo2 (type 316L)	Pipe tensile strength	$\geq$480 MPa	
	Pipe elongation	$\geq$35%	

Table 11.8 Applicable conditions and uses of different brand stainless steel plates

Brand	Applicable conditions	Uses
06Cr19Ni10 (type 304)	Chloride content ≤200 mg/L	Drinking water, drinking cold water, hot water and other pipes
022Cr19Ni10 (type 304L)		Pipes with corrosion resistance requirements higher than type 304
06Cr17Ni12Mo2 (type 316)	Chloride content ≤ 100 mg/L	Pipes with corrosion resistance requirements higher than type 304
022Cr17Ni12Mo2 (type 316L)		Seawater or high chlorine media

3. The applicable conditions and uses of different grades liner with stainless steel plates can be selected according to Table 11.8.
4. The thickness and perimeter of the stainless-steel plates should be determined according to the design requirements. The raw material size of the stainless-steel plates should be selected based on the convenience of construction. The blanking size of the stainless-steel plates used for the preformed pipe blank should be cut according to the design size.
5. The stainless-steel welding consumables used for welding should match the stainless steel plates used.
6. During the transportation, storage and construction of the pipe joints, semi-finished products and structural parts (parts) used, measures shall be taken to prevent their damage and corrosion. Stainless steel materials should be stored in a warehouse away from the pollution of oil, acid, alkali, salt and other chemicals. Prefabricated stainless steel tube blanks should not be stacked to prevent deformation and wrinkling after compression.
7. The negative pressure should be considered in the design of the stainless-steel lining. Special design should be made for places that are prone to negative pressure, such as local high points and downstream of main line valves, to avoid negative pressure causing the plates to collapse.
8. The segment plates should consider the length, nodes, equipment materials and other issues to determine the appropriate length of the segmented plates and the location of the breaking point.

11.3 Equipment Use and Maintenance

11.3.1 The equipment used includes plate rolling machine, argon arc welding machine, pipe support, in-pipe transport vehicle, generator, blower, winch, lighting transformer, etc.

11.3.2 12V AC or DC electrical appliances should be used for the lighting equipment in the tube.

11.3.3 The safe power consumption of the in-pipe welding machine shall comply with the relevant regulations.

11.3.4 Gas cutting tools are strictly prohibited in the pipe.

11.3.5 The equipment needed on site includes conventional equipment and special equipment. The conventional equipment should be used and maintained according to relevant regulations and instructions. The special equipment should be used and maintained according to the equipment instruction manual.

11.4 Construction Preparation

11.4.1 Pipe cleaning and testing shall comply with the relevant provisions of Chaps. 5 and 6 Regulation.

11.4.2 When the pipeline needs to be interrupted, the flow direction of the media in the pipe should be investigated. Corresponding rehabilitation method for interrupting pipe should be adopted according to the original pipe material. The interrupting pipe, the treatment of residual liquid in the pipe and the laying of temporary pipelines should ensure safety and convenient restoration.

11.4.3 The original pipe should be repaired for cracks, staggers, deletions, collapses, etc., and leaks should be blocked. For the misalignment greater than 8% of the pipe diameter, it should be corrected first.

11.4.4 Before the installation of the stainless-steel plates, the inside of the original pipe should be kept tight, leak-free, dry and continuous forced ventilation. The construction personnel in the pipe should wear complete labor protection equipment, and the power cord in the pipe should be well insulated.

11.4.5 The size of the pipe segmental plate should ensure that it can pass through the wellbore.

11.4.6 When the work pit is to be excavated at both ends of the pipe to be repaired, the size of the work pit should be adapted to the length and width of the pipe segment, and the depth of the pit is equal to the bottom of the pipe. The slope excavation, step excavation or support of the work pit wall should meet the requirements of GB50268 and JGJ120.

11.4.7 The inner diameter of the original pipe shall be accurately measured, and the cut size of the stainless-steel plate shall be determined according to the measurement results. For the stainless steel plate of special parts such as elbows, reducing diameters, branch pipes, etc., the size of the lining parts should be accurately measured in advance, and the material should be cut according to the design drawing.

11.5 Pipe Segmental Lining Construction

11.5.1 Module Lining

1. When manually entering the pipeline for construction, the water level in the pipeline should not exceed 30% of the vertical height of the pipeline and no more than 500 mm. Special attention should be paid to the safety of underground personnel and communication between the ground and underground personnel must be maintained.
2. When assembling and connecting the segments, the assembly ring in the inspection well is adopted according to the site conditions. The segments are assembled either from the bottom to the top or from the middle to both sides. The connection of each segment should use a pneumatic tool to ensure that the segment is firmly connected.
3. When bolts are used for connection between the segments, sealant or adhesive should be injected into the connection parts to ensure waterproof performance, and the lock should be rigidly connected.
4. The gap between the new pipe and the original pipe should be grouted and filled, and if necessary, the new pipe should be supported before grouting to ensure the safety of grouting.
5. After the repair construction is completed, the pipeline port should be treated to ensure that the pipe port is smooth and complete.

11.5.2 Stainless Steel Lining

1. Before the stainless-steel plate is fed into the original pipe for welding, the sheet should be rolled into a pipe-shaped billet in advance using special pipe-rolling equipment. The angle and radius of curvature of the pipe should be determined according to the pipe diameter, and the length of the pipe billet should be less than the length of the work pit.
2. Transport the rolled pipe blank to the site, put the pipe blank into the operation pit, and send the pipe blank into the original pipe layout place with a transportation trolley. It is required that the gap between the longitudinal welds of the two sections of pipe blanks should be greater than 200 mm and should be at 4 or 8 o'clock. The lap sequence in the pipe should be in the direction of water flow.
3. Use a special pipe support device to support the stainless steel pipe blank round, spot welding every 100 mm, so that the stainless steel pipe and the inner wall of the original pipe tightly fixed; When supporting the pipe, the pipe blank should be closely attached to the inner wall of the original pipe.
4. Welding in the pipe shall meet the following requirements:

    (a) Stainless steel welding operations shall comply with the relevant provisions of the current national standard "Code for Construction and Acceptance of Field Equipment Industrial Pipe Welding Engineering" GB50236;
    (b) When the high temperature of the welding operation is prone to adversely affect the original pipeline, heat insulation measures should be taken

    (c)   When the butt welds are aligned, the inner wall should be flush, and the amount of misalignment of the inner wall should not be too large;

    (d)   When welding stainless steel, the distance between longitudinal welds should be greater than 200 mm to avoid cross welds;

    (e)   At the end of the original pipeline, a full weld sealing treatment should be carried out between the stainless-steel liner and the original pipeline inner wall.

5. The stainless-steel plate should have a neat appearance, no porosity, no under-penetration, no cracks, no weld lumps, and no overburn. Weld seam quality inspection is strictly performed in accordance with welding quality acceptance standards.

6. After the plate welding installation is completed, the welds in the pipe shall be inspected for flaw detection, and the quality of the welds shall be reliable before the subsequent operations.

7. After the operation of the stainless-steel plate is completed, the corresponding sealing, connection and anti-corrosion treatment should be performed according to the design plan. For the pipeline ports that cannot be connected in time, measures such as blocking and covering should be taken to protect the pipeline ports.

8. For the pipelines with relatively serious defects in the original pipelines, the annular gap between the stainless-steel lining and the original pipeline wall shall be grouted.

11.6   Quality Assurance Measures

11.6.1 If the pipeline is blocked or diverted, a double-insurance blocking wall shall be provided.

11.6.2 Check the certificate, profile specifications, production date and use period of the materials before construction to ensure that the quality of the materials and the specifications of the materials used are consistent with the design.

11.6.3 After welding each section of stainless-steel pipe blanks, perform a visual inspection of the weld seam according to the technical requirements, without porosity, cracks, or burn-through. If a defect is detected, it must be removed with an electric grinder tool. After determining the authenticity, continue to repair. For larger defects, repeated melting methods must not be used to eliminate the defects. Inspection must be carried out on the same day when welding joints completed, if defects found, it should be eliminated on the same day to make sure no water leakage and the pressure bearing capacity shall meet the requirements to be qualified.

11.6.4 In the installation of PVC segments, pay attention to the installation of the sealing device, especially the installation of the last segment, to ensure the sealing into the groove.

11.6.5 In the installation of PVC segments, the operator should pay special attention to the connection of the buckle, and the locking screws should be provided on the sides of the segments.

11.6.6 The design or installation of PVC segments shall resolve the influence of temperature changes and vibration on its stability through wedge-shaped structures and grouting.

11.6.7 The grouting should be carried out in batches according to the designed ratio to ensure that the highest position is grouting.

Chapter 12
Modified Sliplining

12.1   General Provisions

12.1.1 The modified sliplining includes: deformed lining and swagelining. The material used to modified sliplining is a high density polyethylene pipe. The deformed lining is suitable for the renovation of DN50 ~1600 mm pipeline; Swagelining is suitable for the renovation of DN50 ~800 mm pipeline.

12.1.2 When using the modified sliplining to renovate the gas pipeline, the maximum allowable working pressure of the pipeline after renovation should not exceed 0.4 MPa.

12.1.3 Before using the modified sliplining to renovate the pipe, the feasibility assessment of the construction process should be carried out according to the actual situation of the project; if necessary, a pipe section with the same diameter, material and cross-sectional shape as the inner lining pipe and the length not less than 3 m can be used for trial sliplining, and detect the surface damage of the pipe section after trial sliplining, the scratch depth on the outer surface of the pipe should not be greater than 10% of the wall thickness of the liner.

12.2   Construction Design and Materials

12.2.1 The maximum single section length of modified sliplining should not exceed 300 m. When it exceeds 300 m, the pulling force required for construction and the tensile strength of the pipe shall be checked and calculated, and the construction shall be carried out only when the requirements are met.

12.2.2 The working pit shall be excavated at both ends of each pipeline to be renovated, and the length of the initial working pit shall be determined by the following formula:

© China Architecture & Building Press 2021
L. Wang et al., *Technology Standard of Pipe Rehabilitation*,
https://doi.org/10.1007/978-981-33-4984-1_12

$$L = [H \times (4R - H)]^{1/2} \tag{12.1}$$

In the formula:

L Length of starting work pit, m;
H Distance between pipeline center to be renovated and road surface, m;
R Polyethylene pipe is allowed to bend radius, m, and $R \geq 25d_n$;
d_n Outer diameter of pipeline to be renovated.

12.2.3 For renovation of the same pipe section, the inner lining pipe of the same material shall be used. The inner lining pipe shall not have visible cracks, holes, scratches, inclusions and other damage defects.

12.2.4 The electro fusion welding connection should be made before the lining pipe is reduced in diameter. The electro fusion welding connection of the lining pipe should be in accordance with the "Code of Practice for the Preparation of Plastic Pipes and Pipes Polyethylene (PE) Pipes/Pipes or Pipes/Pipes Hot Melt Butt Assemblies" GB19809.

12.2.5 Polyethylene pipes and fittings produced with special blends should be used for lining pipe.

12.2.6 After the lining pipe arrives at the construction site, the mechanical properties of the pipe should be tested according to the production batch, and the qualified pipe can be used for construction.

12.3   Equipment Use and Maintenance

12.3.1 The Equipment used are including:

1. Special equipment for lining: The equipment used for the deformed lining is a U-shaped pressing machine, and the equipment used for swagelining is divided into a die-reducing machine and a cold-drawing machine;
2. Pulling device: composed of winch, guide pulley, wire rope, pulling head, etc. The pulling system shall be equipped with an automatic display device, and the pulling force shall be recorded during construction. The maximum pulling force shall not be greater than the maximum pulling force required by the specification.

12.3.2 The equipment required on site includes conventional equipment and special equipment. The conventional equipment is used and maintained according to relevant regulations and instructions. The special equipment needs to be used and maintained according to the equipment instruction manual.

12.4 Construction Preparation

12.4.1 The cleaning and inspection of the pipeline to be renovated shall meet the requirements of Chaps. 5 and 6 of this regulation.

12.4.2 After the cleaning work is completed, a HDPE pipe with the same diameter as the reduced or folded liner pipe and a length of not less than 3 m should be used for trial wear. Wire rope pulling should be used before and after the trial wear pipe. If there are obstructions or scratches that exceed 10% of the liner wall thickness during the try-on process, the pipeline should be reprocessed and cleaned until the requirements are met.

12.5 Modified Sliplining Construction

12.5.1 Construction for deformed lining.

1. On-site deformed lining construction should be carried out under the conditions of 5–30 °C ambient temperature.
2. A visual inspection should be carried out before the polyethylene lining pipe is pulled in. The degree of scratches on the pipe surface should not exceed 10% of the pipe wall thickness.
3. When pulling into the folded liner, the pulling force should be applied slowly to prevent the pulling force of the folded pipe from exceeding the maximum axial pulling force allowed by the pipe.
4. During the process of pulling into the folded lined pipeline, necessary measures should be taken to prevent the lined pipeline from being scratched by the ramp, the operation pit wall, and the pipeline port to be repaired. Care should be taken to observe the condition of the liner at the entrance of the pipeline to be renovated to avoid excessive bending or wrinkling of the liner.
5. The maximum stretch rate shall not exceed 1.5% when the folded liner is pulled in
6. After the lining pipe is pulled in, the pulling force on the pipeline should be removed, and the two ends of the lining pipe should be reserved for the 300-500 mm reserved section of the pipeline to be repaired.
7. The folding of the lining pipe shall meet the following requirements:

   (a) The pipe should be placed on the ground roller bracket during the folding process, and it is strictly prohibited to drag on the ground;
   (b) After the pipe is folded, it should be tightly wound with winding tape immediately, and the pulling end should be continuously wound;
   (c) The winding distance of the winding belt should be determined according to the diameter and wall thickness of the lining pipe, and the spacing should not be greater than 50 mm;

8. Pulling the lining pipe into the pipeline to be renovated shall meet the following requirements:

   (a) Before pulling in, the introduction device and protection device should be installed at the pipeline starting end;
   (b) The pull-in speed should be controlled at 5–8 m/min;
   (c) The folded lining pipe should have a construction margin that is not less than 1.5 m in the work pit.

9. The restoration of the folded lining pipe should meet the following requirements:

   (a) When the folded lining pipe reach the right place in the pipeline to be renovated, the blind plate should be welded to the end of the folded lining pipe;
   (b) The speed of boosting should be strictly controlled during recovery, and the recovery pressure should be strictly performed according to the parameters of the construction process assessment;
   (c) After the folded lining pipe returns to a round shape and the pressure is stable, the stable time of this pressure should not less than 24 h

10. The recovery process of the factory prefabricated folded lining pipe should meet the following requirements:

   (a) During the recovery process, the changes of the temperature and pressure should be recorded to ensure that the temperature and pressure of each stage of the recovery have reached the relevant requirements. A temperature measuring instrument should be installed between the old pipe and the lining pipe to detect the outside temperature of the lining pipe;
   (b) Inject steam with a temperature of 112–126 °C and a pressure of 100 kPa into the folded lining pipe. When the temperature around the lining pipe reaches 85 ± 5 °C, pressurize the steam pressure to 180 kPa;
   (c) Maintain the steam pressure for a certain time to fully expand the folded pipe and form a concave shape at the branch tube;
   (d) The temperature of the folded lining pipe should be cooled to below 38 °C, and then slowly pressurized to about 228 kPa, and then continue to cool with air or water until the surrounding temperature;
   (e) After the lining pipe is restored and cooled, cut both ends of the pipe neatly. Both ends of the lining pipe should be at least 100 mm longer than the pipe to be renovated.

11. After the factory prefabricated folding pipe returns to round, the water in the pipe should be drained and the pipe should be dried.

12. The electric fusion and hot fusion connection between the folding pipes in the working pit shall meet the following requirements:

   (a) Before connection, stress relaxation of not less than 24 h should be performed, and a fixed point should be set on the polyethylene pipe;

    (b)   When the standard size ratio (SDR) of the lined pipe is 26, the folding pipes in the work pit should be connected with reducing pipe fittings;

    (c)   Before the inner liner is connected, a rigid inner support bush should be installed at the port of the folded pipe;

    (d)   The annular space between the folded inner liner and the steel pipe should be filled with flexible breathable material in the work pit.

13. The lining pipe and the old pipe shall be connected by a steel-plastic conversion joint or a flange-type steel-plastic conversion, and the connection shall comply with the relevant industry regulations.

12.5.2 Swagelining construction.

1. The die shrinking-diameter method should follow the following regulations:

    (a)   The reduction in the diameter of the lining pipe should not exceed 15%;

    (b)   During the pull-in process, a certain amount of pulling force should be applied to the reduced-diameter lining pipe to prevent a large rebound before it completely enters the pipeline to be renovated.

2. The swagelining should follow the following regulations:

    (a)   The reduction in the diameter of the liner should not exceed 15%;

    (b)   It is advisable to preheat the lining pipe to about 100 °C before drawing and reducing the diameter;

    (c)   During the pulling process of the pipe, the pulling force should be evenly applied, and the pulling force should be prevented from exceeding the maximum pulling force the pipeline can bear.

3. The process of reducing the diameter of the liner should be continuous and should not be interrupted.

4. During the drawing process of the lining pipe, the pulling force should be applied slowly and evenly to prevent the tensile stress of the reduced-diameter pipe from exceeding 50% of the yield strength of the material.

5. During the pull-in process, the lining pipe wall should not be worn, scratched, and deformation.

6. During the drawing process of the lining pipe, the stretching rate shall not exceed 1.5%.

7. After the lining pipe is pulled in, the tension should be removed, and a length of 300–500 mm should be reserved at both ends of the lining pipe.

8. After the reduced-diameter pipe is pulled in, the pipe should not stand for less than 24 h to fully restore the original diameter; heating and pressurization can also be used to accelerate the recovery of the reduced-diameter pipe, and the time should not be less than 8 h.

9. The connection between the lining pipe and the old pipe shall comply with relevant industry regulations.

12.6   Quality Assurance Measures

12.6.1 The materials used should conform to the relevant national codes and standards and be re-inspected according to the specifications. Materials that do not meet the specifications are prohibited.

12.6.2 Perform functional tests should be in accordance with relevant technical regulations and acceptance specifications.

Chapter 13
Cured in Place Pipe (CIPP)

13.1 General Provisions

13.1.1 Pipes suitable for CIPP is that whose materials are steel, reinforced concrete, cast iron, plastic, etc.

13.1.2 The cross-sectional shape of the pipeline whoes cross-sectional shape is round, rectangular, oval, etc are applicable for the CIPP.

13.1.3 Generally, the drainage pipe applicable to the CIPP is the pipes whose diameter is 200~2700 mm; the water supply pipe applicable to the CIPP is the pipes whose diameter is 200~1500 mm; the gas pipe applicable to the CIPP is the pipe whose diameter is 200~600 mm; The applicable diameter of UV CIPP is generally φ200~1600 mm.

13.1.4 The maximum length of a single section of pipeline for hot water or steam CIPP should not exceed 300 m; the maximum length of a single section of pipeline for UV CIPP should not exceed 150 m.

13.1.5 The maximum angle of the pipe elbow applicable to the CIPP is 45°; the angle of the pipe elbow applicable to the UV CIPP is 11.25°.

13.1.6 For the renovation of drinking water pipelines, the materials must meet the requirements of the health department.

13.2 Construction Design and Materials

13.2.1 The impregnated hose material of the CIPP is generally composed of thermosetting resin, needle felt or glass fiber.

13.2.2 The impregnated hose used in the CIPP shall meet the following requirements:

© China Architecture & Building Press 2021
L. Wang et al., *Technology Standard of Pipe Rehabilitation*,
https://doi.org/10.1007/978-981-33-4984-1_13

1. The hose should consist of a single layer or multiple layers of needle felt or non-woven or woven material combination, which should be able to impregnate the resin and not react with the resin, and can withstand the tensile force, pressure and curing temperature of the construction;
2. The outer surface of the hose should be covered with a non-permeable plastic film compatible with the resin used;
3. The joints between the layers of the hose must be staggered and must not be overlapped; the tensile strength of the hose is in accordance with "Tensile properties of textile fabrics Part 1 Determination of breaking strength and breaking elongation" GB/T3923.1 The result of test shall not be less than 5 MPa.
4. The hose should be able to stretch and have a certain toughness, and its length should be greater than the length of the pipeline to be renovated; the diameter of the hose should ensure that it can be closely attached to the inner wall of the old pipeline after curing;
5. The supplier shall provide a test report of the initial structural performance of the hose after curing.

13.2.3 The resin used in CIPP method should meet the following requirements.

1. The resin can be thermosetting resin, and a certain number of additives should be added to meet the design requirements such as controlling the curing time;
2. The resin should be able to cure under the action of hot water, hot steam or ultraviolet rays, and the initial curing temperature should be less than 80 °C.

13.2.4 For the resin cured by UV light, the UV light generating device should be able to meet the requirements of the corresponding tube diameter and resin curing.

13.3   Equipment Use and Maintenance

13.3.1 The equipment used includes generators, steam or hot water or UV light generators. If pipes are produced on site, equipment for making impregnated hoses should also be provided.

13.3.2 The equipment required on site includes conventional equipment and special equipment. The conventional equipment is used and maintained according to relevant regulations and instructions. The special equipment needs to be used and maintained according to the equipment instruction manual.

13.4 Construction Preparation

13.4.1 Pipeline cleaning and testing shall comply with the provisions of Chaps. 5 and 6 of this standard.

13.4.2 Pipeline pretreatment should meet the requirements of CIPP process and engineering design.

13.4.3 The amount of resin should be carefully calculated before immersing the hose. The various components of the resin should be fully mixed. The actual amount of preparation should be 5–10% more than the theoretically calculated amount.

13.4.4 After the resin is mixed, it should be impregnated in time, and the residence time should meet the requirements of the manufacturer; if it cannot be impregnated in time, the resin should be refrigerated, and the temperature of the refrigerated should be below 15 °C.

13.4.5 The hose should be impregnated with resin under vacuum, the hose material should be fully saturated with resin, and the number of dry spots or bubbles should be controlled within a reasonable range.

13.4.6 The temperature of the hose should be controlled during storage and transportation of the resin-impregnated hose, and the temperature and time during storage and transportation should be recorded; when the hose is immersed at the construction site, the temperature of the immersion work environment should be lower than 20 °C.

13.5 Construction of Inversion Method

13.5.1 The impregnated hose can be inverted into the pipeline to be renovated by water pressure or air pressure. The inversion pressure should be large enough to allow the impregnated hose to invert to the other end of the pipe and make the hose tightly adhere to the wall of the old pipe; the inversion pressure must not exceed the maximum allowable tension of the hose.

13.5.2 During inversion, the proper inverting speed should be maintained to make the hose stretch smoothly.

13.5.3 During inverting, lubricants can be used to reduce the inversion resistance. The lubricants used must be non-toxic oil-based products. They must not adversely affect resin-impregnated hoses, boilers, and pump systems, and will not breed bacteria and affect the flow of liquid.

13.5.4 After inversion, the resin impregnated hose should be longer than 200 mm at both ends of the original pipe.

13.5.5 Hot water or hot steam can be used to cure the inverted impregnated resin hose.

13.5.6 Curing with hot water should meet the following requirements:

1. The heating rate of the hot water should be controlled according to the pipe diameter, material wall thickness, resin material, type of curing agent and ambient temperature, etc., so that it slowly reaches the temperature required for resin curing;
2. A temperature measuring instrument should be installed to monitor the temperature when water flows in and out;
3. A temperature sensor should be installed between the impregnated resin hose and the old pipe both at the beginning and end of the renovated section to monitor the temperature change of the pipe wall. The temperature sensor should be installed at least 300 mm away from the inside of the old pipe port;
4. It is advisable to determine the curing state of the resin through the resin exotherm curve monitored by the temperature sensor.

13.5.7 The use of hot steam curing should meet the following requirements:

1. The hot steam should be slowly heated up to the temperature required to cure the resin. The temperature and time required for curing should consult the resin material manufacturer;
2. The steam generating device should have a suitable monitor to accurately measure the temperature of the steam, and the temperature during the curing process of the liner pipe should be measured and monitored;
3. We can know about the curing state of the resin through the heat release curve of the resin monitored by the temperature sensor

13.5.8 During the curing process, consideration should be given to the material of the renovated pipe section, the thermal conductivity of the surrounding soil, the ambient temperature, the groundwater level, etc., in order to properly adjust the curing temperature and time. During the curing process, changes in temperature and pressure should be recorded in detail.

13.5.9 After the hose is cured, it should be cooled first, and then the pressure should be reduced. When water cooling is used, the inner liner should be cooled to below 38 °C, and then depressurized; when using air cooling, it should be cooled to below 45 °C before depressurizing. When draining or exhausting pressure, the pressure should be reduced slowly. Vacuum in the pipeline must be prevented from damaging the liner.

13.5.10 The lining pipe end should be cut neatly at the exit of the renovated pipe section. If the lining pipe is not tightly bonded to the old pipe, a resin mixture should be filled between the lining pipe and the old pipe to seal.

13.6 Construction of Pulling Method

13.6.1 Before pulling in the soft liner, a cushion film with a width greater than 1/3 of the circumference of the pipe should be laid on the bottom of the old pipe, and the cushion film should be fixed on both ends of the old pipe.

13.6.2 The pull-in operation of the soft liner shall comply with the following regulations:

1. Before pulling the soft liner into the old pipe, it should be folded in half;
2. The soft liner impregnated with resin should be smoothly and slowly pulled into the old pipeline along the smooth cushion film at the bottom of the pipe, and the pulling speed should not exceed 5 m/min;
3. When pulling in the soft liner, avoid wearing or scratching of the soft liner;
4. The stretch rate of the soft liner shall not exceed 2%;
5. Both ends of the soft liner should be 300–600 mm longer than the old pipe.

13.6.3 The inflation device should be installed at the inlet end of the hose. The gas generating device used should be able to control and display the pressure.

13.6.4 The expansion of the soft liner shall meet the following requirements:

1. Before inflating, you should carefully check whether each connection is well sealed. A pressure regulating valve should be installed at the end of the hose to prevent the air pressure in the pipe from being too high;
2. The air pressure should enable the soft liner to fully expand and close to the inner wall of the old pipe.

13.6.5 Two methods of hot steam or ultraviolet light should be used to cure the resin-impregnated soft liner.

13.6.6 When using hot steam curing, the temperature sensor should be installed on the outer surface of the soft liner at the beginning and end of the old pipe, and the installation position should be at least 300 mm away from the inside of the old pipe port.

13.6.7 Steam curing shall comply with the relevant provisions in 13.5.7 of this standard, and the pressure and temperature of the curing process shall be recorded in real time in detail.

13.6.8 After the curing of the soft liner is completed, cooling and depressurization shall be carried out. The cooling and pressure reduction shall comply with the relevant provisions in 13.5.9 of this standard.

13.6.9 The following requirements should be met when curing with UV light:

1. When the UV lamp is installed, the inner membrane of the soft liner should not be cut;

2.   Consult the material supplier to obtain the best resin curing time and pressure;
3.   The advance speed of the UV lamp and the air pressure in the pipe should be reasonably controlled to ensure that the resin is completely cured;
4.   After the resin is cured, the pressure in the pipe should be slowly reduced;
5.   When the inner diameter of the pipeline to be renovated is less than 600 mm, the ring stiffness value of the liner using UV CIPP repair process should not be less than 5000 N/m^2.

13.7   Quality Assurance Measures

13.7.1 Check the quality certificate, specifications, production date and use period, temporary storage temperature of the hoses before construction to ensure that the quality of the materials and the specifications of the materials used are consistent with the design.

13.7.2 In the process of the material entering the original pipeline, a special person should check whether the hose is damaged; if there are more serious conditions, the on-site professional and technical personnel should be notified in time to take corresponding measures; if the situation is particularly serious, the construction should be stopped. The inversion method should control the inverting speed by controlling the flow of injected water or gas; the UV CIPP method should control the material moving speed by controlling the speed of the hoist.

13.7.3 During the curing process of the inversion method, the operator should pay special attention to the uniformity of hot water or steam, the speed of rising and cooling, and the pressure in the pipe; during the UV CIPP curing process, the operation should be strictly in accordance with the inflation pressure and inflation speed, the walking speed of the UV lamp, and the temperature control required by the supplier.

13.7.4 In case of pipeline plugging and diversion, double-insurance plugging walls shall be provided for pipelines above DN600.

13.7.5 The bundling of the ends of the pipe curing by inversion method should be reliable, and two bandings should be set at each end; the inlet and outlet pipes for hot water curing should be distributed at both ends of the pipe; the steam delivery pipe for steam curing should be located in the center of the pipeline to be repaired.

13.7.6 During the process of putting the UV lamp into the hose, if the hose sags, stop putting the lamp immediately; The connector should be provided with anti-shedding straps; if an abnormal situation occurs during the UV curing process, the emergency stop button of the curing machine and the intake valve should be pressed first, and then other arrangements should be made.

Chapter 14
Spray Lining

14.1 General Provisions

14.1.1 Before pipeline inspection and evaluation, pre-inspection should be taken, and the pipeline condition should be preliminarily judged.

14.1.2 In addition to meeting the requirements of the general scheme, the key conditions such as the basic conditions, structural and functional conditions of the renovated pipeline, environmental conditions, conditions to be met by spraying construction, and sprayed wall thickness should also be described. The time limit for pipeline renovation should be clear before renovation.

14.1.3 The construction plan for repairing pressure pipelines should reflect the contents of pre-processing procedures such as pipe breakage, descaling or descaling, and flushing in a separate chapter. For the application for approval, a special plan should be prepared.

14.1.4 The construction plan for removing rust and dirt should be formulated according to the results of on-site pipeline inspection, and there should be reliable countermeasures for the pipe with much dirt.

14.1.5 When it is necessary to make work pits during construction, the survey, design and construction shall be conducted with reference to the relevant standards of building foundation pits.

14.1.6 For the investigation of the meteorological conditions at the repair site, the temperature should be known or observed; the temperature of the pipeline during construction should not be below 3 °C, and measures should be taken when the construction site encounters rain, wind and sand.

14.1.7 For the survey of the land occupation at the repair site, the occupation of construction equipment and vehicles in different locations during parking should be known. The entry and exit road survey should know about the impact of different

© China Architecture & Building Press 2021
L. Wang et al., *Technology Standard of Pipe Rehabilitation*,
https://doi.org/10.1007/978-981-33-4984-1_14

weather, different periods of vehicles and pedestrian flow on the entry road and construction site.

14.1.8 For the water source survey on the renovation site, it is advisable to survey the location of the existing water source on the site, the distance from the construction site, the amount of water, and the laying environment of the transmission line. For the power supply survey at the renovation site, it is advisable to survey the power, voltage, location of the power supply and the distance between the construction site and the environment of the transmission line, etc.

14.2   Construction Design and Materials

14.2.1 The construction design shall accurately describe the materials, caliber, pipeline structural conditions of the pipeline to be scaled and corroded, the length of the single section of the renovation section and the angle of the pipeline within the section.

14.2.2 The spraying method and the performance of the spraying material should ensure that the flow pressure and deformation of the renovated pipeline and the water quality and sanitation meet the requirements of the design or relevant industry specifications.

14.2.3 Cement mortar spraying and epoxy resin spraying are only used for functional renovation of pipelines. The selection of spray coating wall thickness should be determined by referring to Tables 14.1, 14.2, 14.3 and 14.4 unless the design company has requirements.

14.2.4 The deviation of spray thickness should meet the design requirements.

14.2.5 The amount of spraying material should be determined by formula (14.1) except the design company has specific requirements.

$$Q = (\pi \times D \times t) \times L \times 10^3 + c \tag{14.1}$$

Table 14.1 Selection table of coating thickness for cement mortar sprayed steel pipe

Pipe nominal diameter (mm)	Coating thickness (mm)	
	Mechanical spraying	Manual spraying
500–700	8	–
800–1000	10	–
1100–1500	12	14
1600–1800	14	16
2000–2200	16	17
2400–2600	18	18

Table 14.2 Selection table of coating thickness for cement mortar spray cast iron pipe

Pipe nominal diameter (mm)	Coating thickness (mm)	
	Nominal value	Minimum vale
40–300	3	2.0
350–600	5	3.0
700–1200	6	3.5
1400–2000	9	6.0
2200–2600	12	7.0

Table 14.3 Coating thickness with epoxy resin

Pipe nominal diameter (mm)	Coating thickness (mm)	
	Wet film	Dry film
15–25	≥0.25	≥0.20
32–50	≥0.25	≥0.20
63–100	≥0.32	≥0.25
150–600	≥0.38	≥0.30

Table 14.4 Coating thickness with polyurethane resin

Internal diameter of pipe (mm)	Thickness of functional repair (mm)	Thickness of structural repair (mm)						
		Buried depth of pipeline to be repaired (m)						
		1.5	1.8	2.1	2.4	2.7	3.0	3.3
100	1.2	3.5	3.5	3.5	3.5	3.5	3.5	3.5
150	1.2	3.5	3.5	3.5	3.5	3.5	3.5	3.5
200	1.2	3.5	3.5	3.5	3.5	3.5	3.5	3.5
250	1.2	3.5	3.5	3.5	3.75	4.0	4.0	4.0
300	1.2	4.0	4.0	4.5	4.5	4.75	5.0	5.0
360	1.2	4.5	4.75	5.0	5.25	5.5	5.75	6.0
410	1.2	5.25	5.5	5.75	6.0	6.25	6.5	7.0
460	1.2	6.0	6.25	6.5	6.75	7.0	7.25	7.5
510	1.2	7.0	7.0	7.25	7.5	8.0	8.25	8.5
560	1.2	7.25	7.5	8.0	8.25	–	–	–
610	1.2	8.0	8.25	8.5	–	–	–	–

In the formula:

Q The amount of spraying material, L;
D Inner Diameter of Pipe, m;
t Spraying thickness, m;
L spray pipe length, m;

   c Considering the amount of solidification and loss (*L*), the polyurethane material should be taken $(4-12) \times 10^3$ (*L*) per spraying; The amount of other materials should be specifically determined in combination with the instructions for use of the paint and the spraying equipment, storage status, and environment

14.2.6 The temporary storage time on site of the spraying material should meet the requirement of the operating instruction.

14.3   Equipment Use and Maintenance

14.3.1 The equipment used includes: generator, blower, air compressor, air sprayer or centrifugal sprayer, etc.

14.3.2 The conventional equipment attached to the spraying shall be used and maintained in accordance with the relevant regulations and instructions.

14.3.3 The anti-loosening device shall be installed on the high-pressure air pipe specialized for spraying.

14.3.4 Cleaning of special centrifugal nozzle components for spraying shall meet the following requirements:

1. After spraying, the nozzle assembly should be disassembled before cleaning: the cup, the paint mixing tube after the rotating cup, the mixing head, and the paint delivery short pipe;
2. No need to clean the mixing tube after rotating the cup;
3. The mixing head and the paint delivery pipe should be cleaned multiple times with a separate cleaning pump, after cleaning the mixing head, spray the de-rusting lubricant to the air inlet;
4. The rotating cup should be placed in the diluent and soaked for 24 h, and then cleaned. It is forbidden to wipe the rotating cup with abrasive materials.

14.4   Construction Preparation

14.4.1 When the work pit needs to be excavated at both ends of the original pipeline, the size of the work pit along the axis of the pipeline should not be less than 2.5 m, the width of the pit should not be less than 1.5 m, and the depth of the pit is 0.5 m lower than the bottom of the pipe. Slope excavation, bench excavation or support of the work pit wall should meet the requirements of "Code for Construction and Acceptance of Water Supply and Drainage Pipeline Engineering" GB50268 and "Technical Specifications for Support of Building Foundation Pit" JGJ120.

14.4.2 Survey and treatment of medium flow before pipe breakage, pipe breakage, treatment of residual liquid in the pipe, and laying of temporary pipelines shall ensure safety and ease of recovery.

14.4.3 Before cleaning, it is advisable to check the status of the pipeline inner wall, such as corrosion, damage, blockages, and the condition of branching pipe, and save CCTV image data.

14.4.4 Before spraying, pipeline pretreatment should be carried out. The pretreatment includes: derusting, descaling, cleaning and drying of the pipeline. The pretreatment should meet the following requirements:

1. It is advisable to use a combination of tools and appliances such as winches, grab harrows, steel wire harrows, etc. to carry out pipeline derusting and descaling operations many times;
2. The derusting and descaling of the pipeline should be selected according to the material of the pipeline, the inner diameter of the pipeline, and the degree of scaling.
3. The pipeline cleaning after derusting and descaling should be carried out by high pressure water jet machine or combination cleaning method;
4. Drying the inner wall of the pipe, removing sediments and accumulated water should use a sponge drying ball or absorbent cotton, and the inner wall should be smooth, free of corrosion, no particles and no accumulated water after the pipeline is dried.
5. For large cracks and holes in the pipe wall, pretreatment should be carried out until the pipe can meet requirements of the spray renovation.

14.5 Polyurethane Spraying Construction

14.5.1 Sprayer in place

1. In mechanical spraying, the placement of the spraying machine should be considered in conjunction with the movement of the conveying pipe and the position of the work pit.
2. The air pipe connecting the air compressor and the sprayer should use anti-off parts.
3. The amount of paint added to the sprayer tank should be determined according to the length of the original pipeline.

14.5.2 Sprayer start-up and paint preheating and pressurization

1. The spraying unit should be started according to the instructions. Start the generator matching the sprayer first, then check the operation of each part of the sprayer, then adjust the sprayer to heating mode, and finally start the air compressor.

2. The preheating time of the paint in the spraying machine's storage irrigation shall be determined according to the ambient temperature of the spraying machine's temperature sensor. The preheating operation of the paint should open the heating switch of the storage tank, the mixer of the storage tank and the umbilical pipe.
3. When the air compressor pressurizes the storage tank of the sprayer, the pressure in the tank should be controlled not to exceed 250 Mbar, and operate according to the following procedure: first confirm that all umbilical valves and return valves are open, then slowly open the storage tank vent valve, next adjust the pressure in the tank to 170 Mbar.

14.5.3 Nozzle assembly and connection.

1. The selection of the rotating cup of the spray head should be determined according to the inner diameter of the pipe and the thickness of the spray.
2. When the inner diameter of the pipeline is 75–300 mm, the rotating cup should be installed on the nozzle pulley. When the inner diameter of the pipeline is 350–600 mm, the rotating cup should be installed on the small trailer of the nozzle.
3. Before connecting the nozzle to the pipeline, add a few drops of rust and lubricating reagent to the air inlet of the nozzle motor.
4. After the nozzle is connected, all the actions of the nozzle should be checked. The rotation speed of the cup should be calibrated with a tachometer. After calibration, the air supply port of the nozzle should be closed in time.

14.5.4 Flow check

1. Before spraying, the two-component paint supply pump of the sprayer and its control system should be checked, and the weight of the paint pumped by the sprayer in the same unit time should be checked.
2. Checking the weigh shall be carried out three times continuously, and the control system of the checking equipment shall give responses in all three times.

14.5.5 Umbilical pipe in place

1. The umbilical pipe should be pulled from the ending work pit to the beginning work pit by the hoist to go through the renovated pipeline. After the spraying is finished, the pulley to protect the umbilical canal should be installed at the tail of the pipeline to be renovated.
2. The end of the umbilical pipe connected to the nozzle should be sealed and protected before the hoisting machine rope is pulled, and the connection with the hoisting machine rope should be reliable.
3. The connection and sealing of the nozzle and umbilical pipe should be reliable.

14.5.6 Spraying

1. The spraying start operation should be carried out according to the following steps:

(a) Before spraying, the air supply switch of the rotating cup should be activated to rotate the rotating cup, and then the air supply pressure should be increased to the rated value;

(b) When the two valves at the end of the umbilical pipe are opened and the paint supply countdown ends, the umbilical pipe reel rotation control switch is activated. Control the pressure of the umbilical reel rotary pump during the spraying process.

2. In automatic spraying, the control parameters should be entered according to the equipment prompts.

3. When spraying ends and the spray head moves out of the pipe, stop spraying immediately.

14.6 Cement Mortar Spraying Construction

14.6.1 When mechanical spraying is used, manual spraying may be used for special pipe fittings such as elbows, tees and adjacent gate valves. The combination of mechanical spraying and manual spraying shall adopt a gradual approach.

14.6.2 When the sprayed cement mortar reaches the final setting, it should be immediately watered and cured, and the mortar layer should be kept moist for no less than 7 days.

14.6.3 During the curing period, the holes in the renovated pipe section with open pretreatment should be tightly closed.

14.7 Epoxy Resin Spraying Construction

14.7.1 Epoxy resin spraying should first choose centrifugal spraying or gas spraying according to parameters such as pipe diameter.

14.7.2 Spray paint should be fully stirred and cured for 15 min before spraying.

14.7.3 For equipment spraying, the flow rate of the nozzle and the traveling speed of the spraying vehicle should be controlled.

14.7.4 For multiple spraying, the latter spraying should be carried out after the surface of the previous spraying is dry.

14.7.5 For air spraying operation, the paint should be put into the sprayer first, and then the sprayer should be connected to the air compressor and the air outlet pipe.

14.7.6 Air spraying should be sprayed more than 2 times.

14.7.7 Excess material for gas spraying should be blown out by high-pressure gas.

14.7.8 The first primer spraying of centrifugal spraying should be completed within 1 h after sandblasting and derusting.

14.7.9 The thickness of centrifugal spraying can be achieved by multi-layer spraying.

14.7.10 For curing after the epoxy resin is sprayed, the breeze should be sent into the tube until the coating film is initially hardened; then it can be cured naturally, or warm air can be sent for heating and curing. When the temperature of heating and curing is 25 °C, the curing time should be more than 4 h; when the temperature of heating and curing is 60 °C, the curing time should be more than 3 h.

14.8 Quality Assurance Measures

14.8.1 In the process of derusting and spraying of pipes, the concentration of harmful substances in the air shall not exceed the provisions of GB6514; the dust content in the air shall not exceed the provisions of TJ36.

14.8.2 For polyurethane spraying, build a sun-proof shed at the pit to ensure that the temperature in the pipeline is not greater than 40 °C in summer construction; build a thermal insulation shed at the pit to ensure that the temperature in the pipe is not less than 3 °C during winter construction.

14.8.3 When spraying cement mortar, the slump of the mortar should be controlled at 60–80 mm; when the pipe diameter is less than 1000 mm, it can be increased but not more than 120 mm.

14.8.4 When the ambient temperature is lower than 5 °C or the humidity is higher than 85%, epoxy resin spraying is not suitable.

14.8.5 During the rainy period, windproof and rainproof sheds should be added to all exposed electrical equipment, the number of electrical equipment detections should be increased, the protection of spraying equipment should be strengthened, and anti-skid facilities on site should be added.

14.8.6 The content of spraying records shall meet the requirements of national and local municipal engineering construction technical data.

Chapter 15
Spirally Wound Lining

15.1 General Provisions

15.1.1 The spirally wound lining should choose the manual winding and the mechanical winding according to the specific situation; the mechanical spirally wound lining has two kinds: expansion lining and fixed diameter lining.

15.1.2 The applicable scope of spirally wound lining should meet the following requirements:

1. It is suitable for the renovation of rain and sewage drainage pipes made of ductile iron pipes, reinforced concrete pipes and other synthetic materials;
2. It is suitable for the renovation of large rectangular box culverts and various irregular drainage pipes;
3. The manual winding is suitable for the overall renovation or localized renovation of the drainage pipe with a diameter of 800–000 mm; the expansion lining is suitable for the overall renovation of the drainage pipe with a diameter of 150–800 mm; the fixed diameter lining is suitable for the localized and overall renovation of the drainage pipe of 450–3000 mm;
4. It is suitable for structural defects of pipelines showing deformation, misalignment less than 3 cm, disconnection, leakage, corrosion, and stable pipeline foundation structure, and no obvious change in pipeline line type;
5. It is suitable for the renovation of local trachoma, dewstone, flaking and other diseases on the inner wall of the pipeline;
6. It is suitable for preventive repairs when the pipeline interface is in the sign of leakage or critical state;
7. It is not suitable for the renovation of structural defects such as pipeline foundation breakage, severe pipeline breakage, severe disconnection of inverted pipelines, severe misalignment of pipeline interfaces, and severe deformation of pipeline linearity.

© China Architecture & Building Press 2021
L. Wang et al., *Technology Standard of Pipe Rehabilitation*,
https://doi.org/10.1007/978-981-33-4984-1_15

15.2   Construction Design and Materials

15.2.1 Spiral pipe can choose independent structure pipe and composite pipe according to specific application situation. The independent structure pipe means that the new pipe does not rely on the original pipe at all and bears all the burden alone; the composite pipe means that the spiral pipe bears part of the load, and the other part of the load is borne by the structural grouting between the new and old pipes.

15.2.2 Strip profile

1. The belt-shaped profile is a belt-shaped product of different widths extruded with PE or PVC. The inner layer is smooth and has a certain thickness. It can be fasten tightly with each other by the lock catches.
2. The performance of materials used to manufacture occlusal strips should meet the design requirements. High-strength should choose the type of PVC profile with stainless steel.
3. The strip profile is rolled into a roll on the drum for easy transportation and storage. The production date and length should be printed on the profile to ensure that the material is used within the period of use.

15.2.3 Profiles with Sealing Adhesive

It is divided into two types: The first type comes with a sealing strip and a electro fusion welding strip when it leaves the factory, and it plays a sealing role in the reduced diameter connection of the male and female locks. The other type is to continuously add silica gel adhesive during the winding construction to play a sealing role.

15.2.4 Grouting material of composite pipe

1. When the spirally wound pipe is used as an independent pressure-bearing struc- ture, the gap between the new and the old pipe can be filled without grouting or by choosing ordinary cement with good fluidity.
2. When the spirally wound lining needs to form a composite pipe through grouting to bear the pressure, the cement slurry must meet the following requirements: good bonding strength with the liner and the old pipe, little shrinkage after curing, and good water barrier.
3. The grouting should be full, and the fluidity and setting time of the slurry should meet the requirements of Table 15.1.

Table 15.1 Performance requirements of slurry

Performance	Index
Mobility (mm)	≥ 270
Condensation period	Initial ≥ 4 h, Final ≥ 8 h

15.3 Equipment Use and Maintenance

15.3.1 The equipment used includes generator, blower, air compressor, sealant pump, special winding machine, etc.

15.3.2 The equipment required on site includes conventional equipment and special equipment. Conventional equipment is used and maintained in accordance with relevant regulations and instructions. Special equipment must be used and maintained in accordance with the equipment instruction manual.

15.3.3 Winding machines and other major underground equipment should be mainly composed of stainless steel parts because they work in water or humid environment all year round. After each construction, it should be cleaned in time and applied with anti-rust agent to ensure the normal use of the machine in the future.

15.4 Construction Preparation

15.4.1 Use high-pressure water to remove all garbage, tree roots and other obstacles that may affect the installation of new pipes.

15.4.2 Sewage pipes that need to be rehabilitated should be detected and recorded with closed-circuit television.

15.4.3 All obstacles should be recorded, and if necessary, they should be washed repeatedly.

15.4.4 The position of the branch pipe must be recorded. When the renovation is completed, reopen the branch pipe.

15.5 Spirally Wound Lining Construction

15.5.1 Construction by mechanical winding-expansion method

1. Initial winding of the pipe: under the drive of the machine, the PVC profile is continuously entangled in the winding machine, and the primary and secondary locks on both sides of the profile are interlocked by spiral rotation, thereby forming a continuous seamless newer than the original pipe liner. After the winding of a section of spiral pipe is completed, the secondary lock is pulled off by pulling the embedded steel wire, so that the new liner begins to expand radially until the new liner is tightly attached to the inner wall of the original pipe.

2. During the winding process, the winding machine continuously repeats the following actions:

    (1) Inject the lubricating sealant into the female lock of the main lock (the lubricating sealant plays the role of lubricating in the process of winding and expanding, and the sealing of the formed liner at the end of expansion);
    (2) Entrapped high-tensile pre-embedded steel wire: When this wire is pulled out, the secondary lock will be cut to allow the new tube to expand, but the steel wire is not pulled out during the winding process of the new pipe;
    (3) The strip profile is rolled into a circular liner;
    (4) The expansion of the pipeline is finally formed. After the initial winding is completed, the winding machine stops working; then drill two holes in the new pipe at the end and insert steel bars to prevent the new tube from rotating in the next expansion; after completion, Start the wire drawing equipment and winding machine. As the embedded steel wire is slowly pulled out, the interlocking secondary buckle is cut during the winding forming process, so that the profile slides along the track of the main lock under the drive of the winding machine and continue to expand in the radial direction until the liner at the non-fixed end (winding machine end) is also tightly attached to the original pipe wall.

3. After the expansion of the new pipe is completed, the ends of the new pipe are sealed with polyethylene foam or polyurethane.

15.5.2 Construction by mechanical winding fixed-diameter method

1. Pipe winding: The winding process of the new liner with the fixed diameter method is similar to the expansion method. When the new liner reaches the next working pit (or manhole), the winding forming process stops.
2. Pipe grouting: After the new liner is wound to a fixed size, a certain annulus space may be left between the mother pipe and the new pipe, which must be filled with cement slurry.
3. The selection of cement slurry admixture should meet the requirements of the relevant specifications; the grouting pressure should be determined according to the test of the ring stiffness of the liner pipe.
4. Other construction procedures, such as branch pipe cutting, can be carried out immediately after winding.

15.5.3 Construction by manual winding method

1. Pipe winding: The manual winding method is different from the mechanical winding method. It is manually formed into the pipe and wound to form. It can be wound from one end of the nozzle to the other end of the pipe, or from the middle of the pipe section to both ends.
2. Pipeline grouting: The grouting method and related requirements are basically the same as the mechanical winding fixed-diameter method. The main difference

is that the manual winding method is used to improve the overall strength of the renovated pipe before the winding construction. A steel frame is installed in the annulus space of the old pipe and the winding pipe, and the steel-concrete structure is formed after the cement slurry is poured.

15.6 Quality Assurance Measures

15.6.1 Before each winding construction, check the certificate, profile specifications, production date and use period of the profiles used to ensure that the quality of the materials and the specifications of the materials used are consistent with the design.

15.6.2 During the winding process, a dedicated person should check whether the profile is damaged or bent, and repair small defects in a timely manner; if there are more serious conditions, the professional technicians on site should be notified in time to take corresponding measures; In some particularly serious situations, construction should be stopped to ensure the quality of each winding.

15.6.3 During winding, the operator should pay special attention to the reduced-diameter connection of the male and female locks and the glue injection in the locks.

15.6.4 The grouting shall be carried out in batches according to the designed ratio.

15.6.5 The spirally wound lining should provide various test and evaluation reports of the technology before construction and use. The report includes but is not limited to the chemical resistance report of the profile, the wear resistance report of the profile, and the tightness report of the pipe, compression resistance report of the pipe and strength report of the grouting material, etc.

Chapter 16
Localized Renovation

16.1 General Provisions

16.1.1 Localized renovation includes spot CIPP, stainless steel foam sleeve, pipe segmental lining, grouting, etc.

16.1.2 There are many localized renovation methods for pipelines, which are difficult to generalize. You can refer to related similar processes. For engineering acceptance of localized renovation, you can refer to related similar processes. For difficult reference methods, you can determine the test method and acceptance procedure.

16.2 Spot CIPP

16.2.1 Scope of application

1. Applicable to the localized renovation of rainwater and sewage pipes whose ducting pipes are ductile iron pipes, reinforced concrete pipes and other synthetic materials.
2. The inner diameter of the pipeline should be 200 ~ 1500 mm.
3. Applicable to structural defects of pipelines such as rupture, deformation of less than 10%, dislocation, leakage, misalignment and interface misalignment should be less than or equal to 5 cm, the basic structure of the pipeline is basically stable, the line shape of the pipeline is not significantly changed, and the wall of the pipeline is not solid crisp.
4. Applicable to the interface pretreatment of the overall repair of the pipeline; the room temperature curing type spot CIPP is suitable for the situation of incompletely blocking the pipeline, and the UV spot CIPP is suitable for the situation of completely blocking the pipeline.

© China Architecture & Building Press 2021
L. Wang et al., *Technology Standard of Pipe Rehabilitation*,
https://doi.org/10.1007/978-981-33-4984-1_16

5. It is not suitable for the renovation of structural defects such as pipeline foundation fracture, pipeline rupture, pipeline disconnection in an inverted shape, serious misalignment of pipeline interface, and serious deformation of pipeline line shape.

16.2.2 Construction design and materials

1. The design of the liner thickness of the spot CIPP should be the same as the thickness design of the CIPP.
2. The length of a single lining pipe should not be less than 400 mm. The length of the lining pipe should cover the defect to be renovated, and it should be at least 200 mm longer than the defect before and after.
3. The curing time of the resin at room temperature with the spot CIPP should be 2–4 h, and it should not be less than 1 h; the curing time for the UV-cured resin with the spot CIPP should be 2 min. The curing time can be determined according to the diameter, length and construction conditions of the renovated section.

16.2.3 Equipment use and maintenance

1. The use equipment includes: in addition to the general equipment for dredging and testing, it should also include generators, air compressors, special airbags, curing equipment (UV curing spot CIPP), etc.
2. If you need to clear the roots, in addition to being equipped with a general-purpose high-pressure cleaner, you should also be equipped with a dedicated cleaning head or a pipe robot.

16.2.4 Preparation before renovation construction

1. Preparation process: periscope pre-detection (severe clogging in the pipe) plugging the pipeline—high-pressure water jet dredging and cleaning CCTV inspection to find pipeline defect points to judge whether to use localized renovation construction—localized renovation equipment, personnel, material preparation; QV pre-preparation inspection (if blocking in the pipe is not serious, CCTV can inspect it) CCTV inspection looks for defects in the pipeline to determine whether to use the localized renovation—localized renovation equipment, personnel, and material preparation.
2. When using high-pressure water jet dredging and cleaning, all silt in the pipeline should be removed. For tree roots and other special obstacles that may affect the renovation of the new pipe installation, special cleaning heads or pipeline robots should be equipped for dredging.
3. Redredging and cleaning if necessary.
4. If the test finds voids in the soil around the pipeline, drilling and grouting should be used to strengthen the soil around the pipeline. Before spot curing and renovation on site, the soil around the pipe should be grouted and strengthened. The grouting liquid needs to fill the interior of the soil layer and the gap, to form an anti-seepage curtain, to strength the stability of the soil around the pipe, to

prevent the loss of soil around the pipeline, to improve the bearing capacity of the foundation soil of the pipeline.
5. The airbag selection for localized renovation should consider the consistency with the inner diameter of the pipeline.

16.2.5 Spot CIPP Construction

1. Felt cutting: According to the situation of the pipeline to be renovated, cut the glass fiber felt that matches the size of the pipeline. The length of the felt cloth should be the circumference of the pipe +200 mm to ensure that the felt cloth partially overlaps on the airbag to ensure that the felt cloth can closely adhere to the mother pipe; The width of the felt cloth should ensure that both sides exceed the defect length of the pipeline by more than 10 cm.
2. Resin curing agent mixing: Firstly, mix the appropriate amount of resin and curing agent mixture according to the formula ratio required by the resin supplier. The amount of resin should be enough to fill the fiber hose, and it should be increased by 5–10% than the design amount. Then mix with a stirring device to make the mixed liquid free of foam and record the mixed humidity. A sample shall be kept for each batch of resin mixture during construction for testing the curing performance.
3. Resin penetration: Use an appropriate spatula to evenly apply the resin mixture to the glass fiber felt cloth and fold the felt cloth to the design value. In order to improve the construction quality and reduce the carried air amount, rollers should be used to roll the resin felt cloth.
4. Felt cylinder positioning installation: The resin-impregnated felt cylinder is installed through the airbag. In order to form an insulating layer between the airbag and the pipeline during construction, the polyethylene (PE) protective film is used to bind the airbag first, and then the felt pipe is bound to the airbag with tape to prevent it from sliding or falling. When the airbag is sent into the renovated pipe section, the air pipe should be connected, and the felt pipe should be prevented from touching the inner wall of the pipe (UV type should also put the curing lamp and cable into the pipe).
5. Reset and curing: After the airbag is in place, use an air compressor to pressurize the airbag to inflate. Inflation should be slow and uniform. The pressure at the room temperature curing airbag should be kept within 0.15 MPa (UV control pressure should be halved) to make the felt cylinder tight, stick to the wall. The air pressure needs to be maintained for a certain period of time until it reaches full curing under normal temperature (UV type also needs light).
6. After fully curing, release the airbag pressure, drag it out of the pipeline, and record the curing time and pressure.
7. When multiple lining pipes are used for extended renovation, the curing should start from the downstream, and the length of each section should be no less than 100 mm.

16.2.6 Quality assurance measures

1. Equipment operators should strictly abide by the equipment operation rules.
2. The end of the liner is smoothly joined to the pipe; when the renovated length is extended and multiple lining pipes are used for lap renovation, the lap joint should be smooth and meet the pressure of the upstream port and the downstream port.

16.3   Stainless Steel Foam Sleeve

16.3.1 Scope of application

1. Applicable pipe are reinforced concrete, plastic pipes, ductile iron pipes and other synthetic materials for rainwater, sewage and water supply pipes.
2. Suitable for the renovation of localized damage to drainage pipes with a pipe diameter of 150–1350 mm.
3. Applicable structural defects of pipelines are disjointed and leaking, the basic structure of the pipeline is basically stable, the line shape of the pipeline is not significantly changed, and the wall of the pipeline is solid and not crisp.
4. Suitable for preventive renovation when there is leakage or criticality at the pipeline interface.
5. It is not suitable for the renovation of structural defects such as broken pipeline foundations, inverted joints of pipelines, severe misalignment of pipeline joints, and severe deformation of pipeline lines.

16.3.2 Construction design and materials

1. The stainless-steel foam sleeve is divided into two layers, which are composed of stainless-steel material and a filler containing polyester foam rubber. A stainless-steel foam sleeve with sponge attached to its outside is installed at the leakage point of the pipeline. The sponge absorbs the polyester foam glue slurry after the installation is in place, use an inflation bladder to make it close to the wall of the pipe, and the slurry expands between the stainless steel and the pipe to achieve the purpose of stopping water.
2. The design strength of the stainless-steel foam sleeve needs to ensure and restore the original pipeline design function. The structure of the repaired pipe is improved in strength and chemical resistance. The foam filler can provide structural protection.
3. Stainless steel sleeves should use austenitic stainless steel 304 or stainless steel 316.
4. The foaming agent adopts the polymer chemical grouting plugging material produced by the polymerization chemical reaction of a polyisocyanate and polyether, etc. which has a good effect of stopping the leakage of the concrete structure.

16.3.3 Equipment use and maintenance

1. The equipment used includes generator, blower, hoist, etc.
2. The equipment required for the localized on-site curing renovation includes conventional equipment and special equipment. The conventional equipment is used and maintained according to relevant regulations and instructions, and the special equipment is used and maintained according to the equipment instruction manual.

16.3.4 Preparation before renovation

1. Use high-pressure water to remove all the garbage, tree roots and other materials in the pipeline that may affect the renovation.
2. Sewage pipelines that need to be renovated should be tested and recorded with closed-circuit television.
3. All obstacles are recorded, and repeated cleaning if necessary.
4. Pipeline dredging and plugging: plugging of pipelines, pumping, dredging, poisoning detection and protection—looking for leakage points and breakage points—stopping water and plugging.
5. Borehole grouting and soil reinforcement: Before localized on-site curing and renovation, the soil around the pipe should be grouted and reinforced. The grouting liquid fills the interior and voids of the soil layer to form an anti-seepage curtain to strengthen the stability of the soil around the pipe and improve the pipeline the bearing capacity of the foundation soil.

16.3.5 Renovation construction

1. Put the stainless-steel foam sleeve on the ground outside the rubber airbag with wheels, the innermost is the airbag, the middle layer is the stainless-steel reel, and the outermost layer is the sponge reel coated with styrofoam.
2. Use the paint roller to evenly coat the foam layer on the outermost sponge layer of the foam sleeve and calculate the volume expansion 3–7 times after foaming.
3. Connect the TV camera, rubber air bag and stainless-steel foam sleeve in series. Under the pulling of the cable, the air bag with wheels and the reel enter the pipeline from the inspection well or the operation port.
4. Under the guidance of the TV camera, position the sleeve at the interface to be repaired.
5. Start the air pump to inflate the rubber airbag. The expansion of the airbag is the expansion of the retracted reel and it is close to the pipe wall.
6. When the sleeve is inflated, the positioning card of the stainless-steel sleeve will lock the reel so that it will not rebound after the airbag is deflated.
7. The stainless-steel foam sleeve, sponge styrofoam and cement pipe are glued together, and the styrofoam is consolidated after a few hours.
8. The quality after renovation is checked by the TV equipment to see if there is slurry coming out around the stainless-steel foam sleeve and whether the water leakage point achieves the water-stop effect.

16.3.6 Quality assurance measures

1. Check whether all equipment is operating properly before construction, and list equipment tools.
2. During the installation process, check the renovation points in the video to remove any obstacles that may affect the installation.
3. Make sure the amount of all styrofoam, and properly lock the stainless-steel foam sleeve to ensure the quality of installation.
4. Check by CCTV to determine whether the renovation quality is qualified, check whether the interface is smooth after renovated, whether the buckle is firmly connected, and whether the foaming agent evenly foams, etc.

Chapter 17
Post-processing

17.1 Port Processing

17.1.1 The handling of ports shall comply with this regulation or relevant regulations of related industries.

17.1.2 The renovated pipelines should be in line with the relevant regulations of related industries.

17.2 Site Recovery

17.2.1 After the leak test is qualified, the backfill operation should be carried out in time and the construction site should be cleaned up.

17.2.2 The backfill of the work pit shall comply with the provisions of the current national standard "Code for Construction and Acceptance of Water Supply and Drainage Pipeline Engineering" GB50268.

17.2.3 The restoration of the construction site should not be worse than the condition before construction.

© China Architecture & Building Press 2021
L. Wang et al., *Technology Standard of Pipe Rehabilitation*,
https://doi.org/10.1007/978-981-33-4984-1_17

Chapter 18
Engineering Acceptance

18.1 General Provisions

18.1.1 In addition to meeting the requirements of this regulation, the content, procedures and records of acceptance shall also meet the requirements of the current industry pipeline standards.

18.1.2 The structural or functional test of the pipeline after renovation shall be carried out in accordance with the relevant provisions of the current relevant pipeline standards.

18.1.3 After the replacement and renovation project is completed, the construction company shall first conduct a visual inspection of the rehabilitated pipeline, as well as the pre-inspection of the strength and tightness test, and notify the relevant departments to accept after acceptance.

18.1.4 If the current standards for pipelines in related industries do not mention the structural or functional acceptance of the renovated pipeline, the test conditions, main parameters and acceptance values should be set according to the design requirements of the design book and the purpose of the test; pipelines with similar test conditions, the structural or functional acceptance after renovation should be based on the test results.

18.1.5 The current standard has clear requirements for the main materials for restoration, and its specifications, dimensions, and performance should meet the requirements of the current standard, and the number of on-site samples for the acceptance of the same batch of products should not be less than 1 group.

18.1.6 For the quality acceptance of the renovation that main material is not clearly stipulated in the current industry pipeline standards, the acceptance shall be carried out according to the design requirements or compared the test value of the main parameters of the material with the standard prescribed value in accordance with the

© China Architecture & Building Press 2021
L. Wang et al., *Technology Standard of Pipe Rehabilitation*,
https://doi.org/10.1007/978-981-33-4984-1_18

current standards; For curing, it is advisable to compare the test value of the main parameters after curing with the standard specified value for acceptance.

18.1.7 The acceptance of the polyethylene (PE) pipe for water supply related to the drainage pipe by by spliting and inserting shall be carried out in accordance with GB/T 13663 of the polyethylene (PE) pipe for water supply.

18.1.8 For the acceptance of the quality of the inner surface of the pipeline renovation after renovation that is not mentioned in the current standards of the relevant industry pipelines, the quality of the inner surface of the pipeline after renovation should be checked and accepted in accordance with the requirements of the design or construction plan.

18.1.9 Pipeline pretreatment, renovation and cleaning, and quality of rehabilitated pipeline, should use CCTV equipment for pipeline inspection according to this regulation and the relevant provisions of the current national relevant industry standards, when the pipe diameter is not less than 800 mm, you can also send a person into the pipeline for inspection and take the image.

18.1.10 The acceptance of the pretreatment and cleaning procedures involving the pipeline to be renovated shall be carried out in accordance with the requirements of the design or construction plan for pretreatment and cleaning.

18.2   Pipeline Pretreatment and Cleaning Acceptance

18.2.1 Main control project: The quality of pipeline pretreatment and renovation cleaning should meet the requirements of Table 5.2. The pretreatment parts should have no influence on the defects of the next repair and renovate process.

Inspection method: Check the CCTV records (or visual records in the tube). For the pre-treatment related to the pipeline functional or structural performance, if it cannot be judged that the pre-treatment can completely eradicate the functional or structural defects, it should be determined through field test verification.

18.2.2 General items: linear and smooth filling, staggered and broken parts of the pipeline, smooth interface, smooth transition of special parts. Before closed renovation (CPP, swagelining, deformed lining, etc.) and interval renovation (sliplining, pipe segmental lining, spirally wound lining, etc., including expansion cracking method), the original pipeline leakage and residual liquid should not affect the construction of subsequent processes.

Inspection method: Check the pipeline pretreatment construction records, materials and physical construction inspection records or reports.

18.3 Acceptance for Pipe Rehabilitation

18.3.1 Quality acceptance for pipe rehabilitation requirements

1. For the spraying lining, CIPP, spot CIPP, stainless steel sleeve, improved swagelining, the allowable deviation of the liner wall thickness shall conform to: the average wall thickness shall not be less than the design value, and the thickness of any point should not be less than 90% of the design value.
2. For the sliplining, deformed lining, pipe cracking, pipe segmental lining, mechanical spirally wound lining method, etc., the liner wall thickness should meet the current national standards or design requirements.
3. The quality of the PE pipe's electro fusion welding butt joints shall comply with the provisions of "Preparation of plastic pipes and fittings polyethylene (PE) pipes/pipes or pipes/pipe fittings and electro fusion welding butt assemblies" GB/T19809. The quality of PE pipe connection shall meet the design requirements.
4. The quality of the inner surface of the lining pipe shall meet the following requirements:

   (a) The inner surface of the lining pipe should be smooth, free from cracks, holes, tensile deformation zones and weak zones, which could seriously affect the structure and function of the pipeline;
   (b) The depth of local scratches on the inner surface of the lining pipe is not more than 2 mm, the length of each scratch is not more than 3 m, and the length of each continuous 50 m extension is not more than 3;
   (c) The uplift of the inner surface of the lining pipe and the relative height of the bubbles do not exceed 5% of the inner diameter of the pipeline to be repaired (the smaller diameter value of the special-shaped pipe), and the local uplift and the number of bubbles are not more than 3 for every 50 meters of continuous extension;
   (d) The maximum depth of the depression on the inner surface of the liner is not more than 3 mm or 50% of the average wall thickness of the liner, and there are no more than 3 local depressions for every 50 m of continuous extension;
   (e) The minimum diameter of dry spots on the inner surface of the liner is not more than 80 mm, and the number of dry spots is not more than 3 for every 50 m of continuous extension;
   (f) When the straight section or the radius of curvature of the pipeline to be repaired is greater than 10 times the inner diameter of the pipe, the fold height of the inner surface of the liner pipe is not more than 6 mm or not more than 2% of the inner diameter of the pipe to be repaired (the special-shaped pipe takes a smaller inner diameter value), and each the folds of 50 consecutive meters of continuous length are not more than 10 places; when the bending radius of the pipeline to be repaired is 5 times (inclusive) ~10 times (inclusive) of the inner diameter of the pipeline, the fold height

of the inner surface of the liner pipe is not greater than 20 mm or not larger than the inner diameter of the pipeline to be repaired (The special-shaped pipe takes a smaller inner diameter value) 3%, and the fold is not more than 10 for every 50 m of continuous extension;

(g) When using the pipe segmental lining, the stainless-steel sleeve and the spot CIPP, the lining pipe should be closely attached to the original pipeline, the lining should be complete, and the lap should be smooth and firm;

(h) When the mechanical spirally wound lining is adopted, the joints of the seams are tightly fitted and the connection is firm, without obvious protrusions, depressions, staggers and other phenomena. It is strictly forbidden to have longitudinal bulges, flattened loops, and seam detachment;

(i) The verticality deviation of the lining pipe ports. If the outer diameter of the lining pipe is less than or equal to DN600, it should not be greater than 4 mm; if the outer diameter of the lining pipe is greater than DN600, it should not be greater than 6 mm.

5. The pipeline after renovation should be free of obvious wet stains and water seepage, and dripping and line leakage are strictly prohibited.

6. If the lining pipe and the pipeline to be renovated are tightly fitted, there should be no leakage between the two layers of pipes; when the sealing device is installed, it should meet the design requirements and the sealing is good; The gap filling should be free of looseness, voids and other phenomena.

18.3.2 The acceptance of the main control items of each rehabilitation method shall meet the requirements of Table 18.1.

18.3.3 General items: The acceptance of the general items of each repairing and rehabilitate method should meet the requirements of Table 18.2.

18.4   Test After Pipe Rehabilitation

18.4.1 The temperature of the lining pipe should not be higher than the temperature of the surrounding soil before the structural or functional test of the pipe.

18.4.2 After the pipeline renovation is completed, the gravity flow pipeline shall be tested for pipeline tightness, and the test shall be conducted in accordance with the relevant provisions of relevant industry standards. Under conditional conditions, selectively conduct well-to-well tightness tests.

18.4.3 After the pipeline renovation is completed, the pressure pipeline shall be subjected to a pressure test. The pipeline shall be tested in stages and boosted in stages according to the actual working pressure requirements. The test shall be carried out in accordance with the relevant regulations of relevant industry standards.

Table 18.1 Acceptance chapter 9 of the main control project of each rehabilitation method

Rehabilitation method	Items and methods of each construction method inspection	
	Inspection items and requirements	Inspection method
Pipe cracking	The entry inspection of new pipes, meets the requirements in Sects. 9.2 and 18.3.1 of this Regulation	On-site measurement or sampling test, inspection of factory certificate, performance test report, sanitary approval documents and supplier's product instructions, etc.
	New pipe connection meets the requirements of Article 18.3.3 of this regulation	Test welding, check the quality of the weld, check the connection record
Sliplining (Discrete sliplining included)	The entry inspection of liner meets the requirements in Sects. 10.2 and 18.3.1 of this regulation	On-site measurement or sampling test, inspection of factory certificate, performance test report, sanitary approval documents and supplier's product instructions, etc.
	Liner connection meets the requirements of Article 18.3.3 of this regulation	Test welding, check the quality of the weld, check the connection record
	Grouting meets the requirements of Sect. 10.5.5 of this Regulation	Check the grouting record
Pipe segmental lining	The entry inspection of segments meets the requirements in Sects. 11.2 and 18.3.1 of this Regulation	On-site measurement or sampling test, inspection of factory certificate, performance test report, sanitary approval documents and supplier's product instructions, etc.
	Lining performance meets the requirements in Sect. 18.3.1	CCTV test or manual test, check test record
	Welded seam with stainless steel inner lining, qualified for flaw detection, reliable	Check the weld flaw detection record
Modified sliplining (Swagelining and folded lining included)	The entry inspection of liner meets the requirements in Sects. 10.2 and 18.3.1 of this Regulation	On-site measurement or sampling test, inspection of factory certificate, performance test report, sanitary approval documents and supplier's product instructions, etc.

(continued)

Table 18.1 (continued)

Rehabilitation method	Items and methods of each construction method inspection	
	Inspection items and requirements	Inspection method
	Folded liner and reduced liner recovery, liner performance, meet the requirements of 18.3.1 of this regulation	CCTV test or manual test, check test record
Cured in Place Pipe (CIPP)	The entry inspection of liner hose and raw materials meets the requirements in Sects. 13.2 and 18.3.1 of this Regulation	On-site measurement or sampling test, inspection of factory certificate, performance test report, sanitary approval documents and supplier's product instructions, etc.
	Lining performance meets the requirements in Sect. 18.3.1	On-site measurement or sampling test, inspection test report and sampling test report
Spraying lining	The entry inspection of liner coatings meets the requirements in Sect. 14.2 and 18.3.1 of this regulation	On-site measurement or sampling test, inspection of factory certificate, performance test report, sanitary approval documents and supplier's product instructions, etc.
	Lining performance meets the requirements in Sect. 18.3.1	On-site measurement and CCTV inspection, review inspection report and CCTV inspection report
	The surface of the inner liner of the liquid epoxy coating should be smooth and without bubbles, scratches, etc., and the wet film should have no flow phenomenon	Observe and check construction records
	The compressive strength of cement mortar meets the design requirements and is not less than 30 MPa. The test method refers to the current industry standard "Standard for Basic Performance Test Methods of Construction Mortar" JGJ/T70	Check the mortar mix ratio and compressive strength test block report

(continued)

Table 18.1 (continued)

Rehabilitation method	Items and methods of each construction method inspection	
	Inspection items and requirements	Inspection method
Spirally wound lining	The entry inspection of liner profiles, sealing adhesives, and inspection grouting materials, meets the requirements in Articles 15.2 and 18.3.1 of this Regulation	On-site measurement or sampling test, inspection of factory certificate, performance test report, sanitary approval documents and supplier's product instructions, etc.
	Lining performance, meets the requirements in Sect. 18.3.1	CCTV inspection, review CCTV inspection report
Localized renovation	The entry inspection of liner materials, meets the requirements in Articles 16.2.2, 16.3.2 and 18.3.1 of this Regulation	On-site measurement or sampling test, inspection of factory certificate, performance test report, sanitary approval documents and supplier's product instructions, etc.
	The installation position of the lining layer is accurate, completely covering the part defects to be repaired and closely fitting the original pipeline. The performance of the lining layer meets the requirements of 18.3.1 of this regulation	CCTV inspection, review CCTV inspection report

Table 18.2 Acceptance of general items for each rehabilitation method

Renovate and rehabilitate method	Check items and methods of each process	
	Inspection items and requirements	Inspection method
Pipe cracking	The pipeline is linear and smooth, the interface is smooth, and the transition of special parts is smooth	Observe, review site inspection records, CCTV inspection records, etc.
Sliplining (Discrete sliplining included)	The pipeline is linear and smooth, the interface is smooth, and the transition of special parts is smooth	Observe, review site inspection records, CCTV inspection records, etc.
Pipe segmental lining	The pipeline is linear and smooth, the interface is smooth, and the transition of special parts is smooth	Observe, review site inspection records, CCTV inspection records, etc.
Improved sliplining (Swagelining and folded lining included)	The pipeline is linear and smooth, the interface is smooth, and the transition of special parts is smooth	Observe, review site inspection records, CCTV inspection records, etc.
Cured in Place Pipe (CIPP)	The pipeline is linear and smooth, the interface is smooth, and the transition of special parts is smooth	Observe, review site inspection records, CCTV inspection records, etc.
Spraying lining	The allowable deviation of the thickness and surface defects of the main cement mortar lining meets the design requirements	Run the test According to the requirements of Article 5.10.3, paragraph 4 of GB50268
	Liquid epoxy coating inner liner thickness and electric spark test	Run the test According to the requirements of Article 5.10.3, paragraph 5 of GB50268
Spirally wound lining	The pipeline is linear and smooth, the interface is smooth, and the transition of special parts is smooth	Observe, review site inspection records, CCTV inspection records, etc.
Localized renovation	Smooth pipeline interface and smooth transition at the boundary	Observe, review site inspection records, CCTV inspection records, etc.

18.4.4 For pipes that are localized renovated, no air-sealing or water-sealing tests are necessary, and the leakage should be judged by CCTV testing. There should be no leakage on the inner wall of the lining pipe. The defect to be renovated should be completely covered. The lining pipe should be in close contact with the old pipe wall and there should be no leakage.

18.4.5 Sampling inspection should verify the data of its process inspection and acceptance according to different renovation processes, and the repair pipeline that meets the design and construction requirements can be tested for strength.

18.4.6 Pipe restoration test, sample tube strength test, peel strength test, hydraulic burst test, etc., can be determined according to different renovation process requirements, relevant industry standard requirements, and based on existing test conditions.

18.4.7 Before construction, if the test conclusion is used as the basis for structural or functional acceptance, the test purpose, main parameters, qualified values, and test methods specified in the design document or construction plan shall be used.

18.5   Completion Acceptance of the Project

18.5.1 The conclusion of project quality acceptance is only qualified and unqualified. Unqualified items should be renovated and reworked to pass.

18.5.2 Project quality acceptance shall be carried out according to sub-items and sub-projects, and shall meet the following requirements:

1. The divisional project can be divided into several divisions according to the length of the pipeline or the re-segment, when the project is small, it may not be divided;
2. The sub-projects are divided by pipeline pretreatment, cleaning and pipeline rehabilitation.

18.5.3 The acceptance judgment for completion acceptance shall meet the following requirements:

1. The sub-project meets the following requirements to be qualified:
   (a) The pass rate of the main control project is 100%;
   (b) The pass rate of general projects reaches 80%, and the maximum deviation is less than 1.5 times of the allowable deviation.
2. All the sub-items of the divisional project shall be qualified, then the divisional project shall be qualified.

Chapter 19
Health, Safety, Environmental Protection (HSE)

19.1 Health Management

19.1.1 Labor protection.
The construction department shall equip employees with corresponding labor protection articles in accordance with national and local government labor protection regulations and standards.

19.1.2 Healthcare.

1. Medical
   The construction company shall prepare medical kits and equip them with first-aid medicines according to the construction area, seasons and operating characteristics.
2. Health Care
   The construction company shall establish the necessary health care system in the aspects of employee health check, disease prevention, food hygiene, etc. and implement it conscientiously.
3. Public health

   (a) Paying attention to the environmental sanitation of the station and clean up the garbage regularly; the dormitory must be disinfected to keep it clean and tidy, and there should armed with anti-rodents, anti-flies and anti-mosquito measures.
   (b) The dining room and kitchen should be kept clean and tidy, the tableware must be sterilized, and unknown and spoiled food should not be consumed.
   (c) Maintain personal hygiene, bathe and change clothes frequently to prevent epidemics.

© China Architecture & Building Press 2021
L. Wang et al., *Technology Standard of Pipe Rehabilitation*,
https://doi.org/10.1007/978-981-33-4984-1_19

19.2 Security Management

19.2.1 Basic requirements.

1. Construction personnel should survey the site carefully, understand the various underground facilities, pipeline distribution and surrounding environment on the construction site, and formulate targeted safety technical measures.
2. Before construction, the construction company shall notify the relevant pipeline use and management units along the route, explain to them the construction plan and the protection measures for the existing pipeline; at the same time, they shall make corresponding emergency plans according to the nature of the pipeline.
3. The construction company shall establish the rules and regulations to ensure safe production, and implement them, and keep records of safety activities.
4. The construction company shall set up a full-time or part-time safety officer. The safety officer shall undergo safety training and pass the assessment.
5. The construction company shall carry out vocational training on safety in production for the employees on duty. Special types of work must be certified to work, and regular site safety inspections should be conducted to eliminate hidden dangers. Carry out education on production safety and accident rescue.
6. The construction company should pay full attention to the natural environment in the construction area to prevent damage to life and property which are caused by natural disasters such as floods, wildfires, and landslides.
7. If the temperature is higher than 38 °C or lower than −30 °C, construction should be stopped and protective measures should be carried out.
8. No alcohol is allowed before or during work. When entering the construction site to work, you must wear fitted work clothes and work shoes, and wear a safety helmet. Do not work shirtless, barefoot or wearing slippers.
9. It is strictly prohibited to store toxic and corrosive chemicals for a long time in the venue. When you need to use them, you must wear protective equipment in accordance with relevant regulations. When working in a work area with radioactive materials, active and effective protective measures must be taken in accordance with the provisions of the Regulations on Radiation Protection and Environmental Protection of Uranium Mine Geological Exploration GB15848.

19.2.2 Construction site safety regulations.

1. Power distribution box and lighting distribution box should be set separately.
2. The mobile power distribution box and switch box should be installed on the fixed bracket, and have measures to prevent moisture, rain and sun.
3. The electrical equipment at the site shall be protected to zero or grounded according to the requirements of the power supply system. Ground resistance should be less than 4Ω.
4. Explosion-proof lamps and lanterns should be used for the lighting of the site; when the electrical equipment is being repaired, the power supply should be cut

off, and warning signs should be hung or a special person should be set up for supervision.

5. Electricity for construction strictly abides by the regulations of "Safety Code for Electricity Supply and Power on Construction Site".
6. Hand-held power tools shall comply with the provisions of "Safety Technical Regulations for Management, Use, Inspection and Maintenance of Hand-held Power Tools" GB/T3787.

19.2.3 Regulations on Flood Control.

1. Try to avoid construction in areas prone to landslides, collapses and debris flow.
2. Try to avoid the flood season or avoid construction in areas that may be affected by the flood. When construction is necessary, drainage ditches and dikes should be dug well.
3. In the flood season, materials and equipment must be stored above the flood level warning line.
4. Construction machinery and electrical equipment should have rain and flood prevention facilities.
5. The roads on the construction site should be unblocked, the drainage system should be good, the site should be clean and tidy, and the loess should not be exposed to the open air.
6. The material should be placed neatly and securely without affecting the fire-fighting equipment, public utilities ground facilities and drainage of its own projects.

19.2.4 Cold protection regulations.

1. During the cold season construction, the premises must be tightly enclosed and equipped with heating facilities.
2. The main water supply pipeline must be wrapped and buried with thermal insulation material. In addition to the bandage of the temporary pipeline, a water discharge valve must be installed in a low-lying place. When the water supply is stopped, the water accumulated in the pipe should be drained.
3. Remove the ice and snow in and out of the site in time, and take anti-skid measures around the site.
4. When the diesel engine, water pump and other equipment are temporarily disabled, the accumulated water must be drained to prevent the machine from freezing and cracking.

19.2.5 Fire prevention regulations.

1. A certain number of fire extinguishers, sand boxes, shovel and other fire extinguishers should be provided in the construction site, and they are not allowed to be used for other purposes.
2. In addition to the weeds around the net house, the width of the fire path should be greater than 5 m. When constructing in forest areas and grassland areas,

preventive measures should be taken in accordance with local fire prevention regulations.

3. When using an open flame for heating in the venue, there must be an appropriate safe distance from the roof, siding, and tower cover. The furnace base must be covered with masonry or heat insulation panels.

4. The exhaust pipe of the internal combustion engine and the chimney of the heating stove shall take into account the seasonal wind direction, and protrude from the appropriate side by more than 0.5 m outside the field. The heat shield fire shield shall be installed at the contact with the field.

5. Always pay attention to the burning situation of the heating stove, and do not allow the flame to go outside, or use oil to support combustion. Unburnt ash shall not be dumped casually. Personnel must completely extinguish the fire when they evacuate the scene.

6. No cigarette butts are allowed to be littered in the 6 rooms, smoking is prohibited in the construction site, and lighting with open flames is prohibited.

7. The oil and other flammable materials stored in the construction site must be properly kept, and fireworks are strictly prohibited. When the lubricant is preheated, it must be supervised by a special person. It is strictly forbidden to directly bake the bottom shell of the diesel engine with an open flame.

8. When the oil is on fire, use fire extinguishers and sand to extinguish it. When an electrical appliance catches fire, you should first cut off the power and then go to the rescue.

19.2.6 Safety regulations for electric welding.

1. Use the welding machine strictly in accordance with the data marked on the nameplate of the welding machine, and do not overload it.

2. Before using the welding machine, you should check that the welding machine is connected correctly, the current range meets the requirements, the shell is reliably grounded, and there is no foreign object in the welding machine before it can be closed.

3. While working, the core of the welding machine should not have strong vibration, and the screws that press the core should be tightened. The temperature of the welding machine and current regulator should not exceed 60 °C during operation.

4. Strengthen the maintenance work, keep the welding machine clean inside and outside, and ensure that the welding machine and the welding cord are well insulated. If there is any damage or burn, it should be repaired immediately.

5. Regularly check the technical status of the welding machine circuit and the insulation performance of the welding machine by an electrician. If there are any problems, they should be eliminated in time.

6. During argon arc welding, the intensity of ultraviolet rays is very large, which can easily cause electro-optic ophthalmia and arc burns. At the same time, ozone and nitrogen oxide compounds are generated to stimulate the respiratory tract. Therefore, the construction personnel must wear all kinds of labor protection

products, such as: white canvas overalls, masks, face masks, protective gloves, foot covers, etc. when operating.

7. When argon arc welding personnel leave the workplace or the welding machine is not in use, they must cut off the power supply. If the welding machine fails, it should be repaired by professionals, and safety measures such as protection against electric shock should be taken during the repair.

8. In the welding and cutting workplaces, there must be fire prevention equipment, such as fire hydrants, fire extinguishers, sand boxes and buckets filled with water.

9. When welding in the pipeline, the container must have inlets and outlets, and be equipped with ventilation equipment; the lighting voltage in the container must not exceed 12 V, and welding or cutting in containers that have been sprayed with paint, rubber, fuel, etc. is strictly prohibited.

10. When conducting electric welding operations in a fire-prohibited place, a fire permit must be obtained, and guardians and fire prevention measures are required before operation.

11. Construction workers should beware of electric shock during construction. Workers should wear insulated rubber shoes and be careful not to be injured by arcs and metal splashes.

12. After the welding or cutting work is completed, carefully check the surroundings of the welding site and confirm that there is no fire hazard before leaving the site.

19.2.7 Safety regulations for gas welding.

1. Before operation, you must wear work clothes, work caps and gloves as required to prevent arc damage and prevent burns.

2. Oxygen cylinders filled with oxygen should be lifted gently and slowly, and the mouth of the bottle must not be lifted. Oxygen cylinders should not be exposed to sunlight or close to high temperature, and should be kept away from fire and flammable materials.

3. The distance between the acetylene gas cylinder, the oxygen cylinder and the welding object is not less than 5 m; the acetylene tube and the oxygen tube cannot be used interchangeably; the tube and the joint must be firmly connected.

4. Turn on oxygen first and then a small amount of acetylene gas when igniting; turn off acetylene gas first and then turn off oxygen when extinguishing. When tempering occurs, acetylene gas should be turned off first, then oxygen.

5. When the welding is suspended, the gas valve of the acetylene gas cylinder and the main valve of the oxygen cylinder must be turned off.

19.2.8 Safety regulations during construction.

1. General safety regulations

    (a) The layout of the construction site should be set according to the location designated by the relevant department, and the site and road cannot be arbitrarily occupied.

(b) When constructing on streets, highways, railways and rivers, operators should wear traffic vests and safety helmets.

(c) When the construction site is in the urban area, eye-catching safety signs, guardrails, warning lights, etc. should be set up and safety officers should be set up. At the same time, the enclosure must be set up, and the warning lights should be set on the enclosure. Irrelevant personnel are prohibited from entering the construction site during the construction process.

(d) When constructing on the roadside, it is necessary to set up obvious traffic road signs and designate a person to coordinate traffic safety.

(e) It is strictly forbidden for non-operators to move mechanical equipment and electrical facilities without permission.

(f) Test and inspect the construction machinery and electrical equipment before construction.

2. Main safety regulations during construction

(a) The operation of the pit, pipe laying, and bracing of the pipe blank should be directed by a special person. In the case of unexpected impact or impact, the necessary precautions should be taken to protect the staff and pedestrians.

(b) The work in the pipeline is a limited space operation. The relevant requirements of the limited space operation should be strictly carried out. The harmful gas tester should be equipped, and the ventilation equipment should be installed.

(c) Personnel shall be forcedly ventilated before entering the pipeline. The ventilation time shall not be less than 30 min. They shall enter the pipeline after passing the gas test. The personnel shall ensure continuous forced ventilation in the pipeline during the operation in the pipeline and carry the oxygen meter and oxygen in the pipeline. Bags to avoid poisoning by harmful gases.

(d) When manually entering the pipeline for construction, the water level in the drainage pipe shall not exceed 30% of the vertical height of the pipeline and not more than 500 mm. Special attention should be paid to the safety of underground personnel. The underground personnel must wear safety belts. There is a person on the ground, it is necessary to have a person responsible for communicating with the personnel working underground. Middle

(e) When derusting and spraying pipes, wear protective clothing, special protective gloves, protective masks, earplugs and goggles.

(f) In the event of an electric shock alarm, the equipment operator shall remain calm, stand in a safe position and remain motionless. It is forbidden to contact other objects, personnel or other objects on the pit wall. Inform the ground commander to contact the relevant section to cut off the power. Get out of the equipment through the dry non-metallic human ladder and go to the ground.

(g) Regularly check cables and electrical equipment to prevent electrical safety accidents.

(h) Observe all other safety precautions.

19.3 Environmental Protection Management

19.3.1 Before construction, the hydrogeology, vegetation, climatic characteristics, human environment, etc. around the construction site should be investigated to understand the environmental management methods of the relevant local authorities, environmental function zone division standards, and pollutant discharge standards, and take necessary measures accordingly.

19.3.2 Protect the land resources and make full use of land resources. Use existing roads as much as possible, road construction must not block and fill drainage channels; construction sites should avoid or reduce the occupation of cultivated land, farmland, forest belts. After completion, the occupied farmland, cultivated land and vegetation shall be restored.

19.3.3 Paying attention to the dealing process for waste gas, waste water and waste residues. Construct a waste liquid pond at a low place on the construction site, introduce construction machinery waste liquid and cleaning wastewater into the waste liquid pond, and then carry out environmental protection treatment.

19.3.4 The use of chemical treatment agents that pollute the environment is prohibited near rivers, lakes or residential areas. Land contaminated with oil and chemical treatment agents should be properly replaced or restored.

19.3.5 The equipment is installed firmly to reduce noise. When the equivalent sound level of construction noise exceeds 70 dB, noise reduction measures shall be taken.

19.3.6 In the process of pipeline derusting and spraying, the noise of various production equipment should comply with the relevant provisions of the Industrial Enterprise Noise Control Design Code GBJ87 Industrial Enterprise Noise Control Design Code.

19.3.7 To protect the ecological environment for working and living, do not destroy the green vegetation, and hunt wild animals.

Chapter 20
Production Management and Technical Archives

20.1 Production Management

20.1.1 General provisions

1. Before construction, the design and/or construction company shall prepare the "Pipeline Renewal Construction Organization Design".
2. Strictly organize the implementation and management in accordance with the design requirements of Chap. 4 Engineering Construction Organization of this Code.

20.1.2 Internal cooperation in the management system

1. The construction company is allowed to formulate corresponding implementation rules or supplementary regulations according to the specific situation.
2. Regarding on-site management, the construction company may refer to the relevant provisions in the engineering construction and municipal construction regulations, and formulate a corresponding management system in accordance with the actual situation of the project.

20.1.3 Equipment installation quality acceptance system

1. In the pipeline rehabilitation construction, when the equipment used involves a part of the project that is more dangerous, it should go through an acceptance team composed of personnel from measurement, safety technology, machinery, installation and supervision. It can be constructed only afterwards.
2. The installation quality of the equipment does not meet the requirements, construction shall not be started, and a dedicated person shall be appointed to rework.

© China Architecture & Building Press 2021
L. Wang et al., *Technology Standard of Pipe Rehabilitation*,
https://doi.org/10.1007/978-981-33-4984-1_20

20.1.4 Post responsibility system

1. The post setting can be set by the construction unit according to the project scale, the process and equipment used, etc.
2. Construction personnel engaged in pipeline rehabilitation shall pass the necessary safety training.

20.1.5 Completion inspection and acceptance system

1. After completion of the pipeline rehabilitation construction, a quality acceptance committee or group shall be formed by the owner, design, supervision and construction departments to timely evaluate and accept the project quality.
2. The main basis of the project quality acceptance standard is the pipeline design document or contract indicators.
3. The "Project Quality Acceptance Report" should be filled in during acceptance.

20.2   Technical Archive

20.2.1 Content of technical files

The acceptance of the completion file of the pipeline rehabilitation project shall include the following:

1. Approval documents for start of construction;
2. Survey data along the project;
3. Old pipeline diagram and data;
4. CCTV and pipeline quality evaluation data after cleaning the inner wall of the old pipeline before rehabilitation;
5. Construction organization design (construction drawing);
6. Qualification certificates and quality guarantees for materials such as pipes and fittings;
7. Construction process and inspection records;
8. Pipeline inspection and evaluation report (CCTV inspection record) after pipeline rehabilitation;
9. Data for handling quality accidents and production safety accidents;
10. Pipeline functional test, penetration measurement and other testing data;
11. Construction summary report of construction, supervision, design and inspection units;
12. Project completion drawing and completion report.

20.2.2 Requirements for technical files

The original records of the pipeline renovation construction and the filling of the reported data shall meet the following requirements:

1. Objective, true, accurate and complete;

2. Reflect professional characteristics, use professional terminology, use professional, standardized and rigorous words;
3. The font is neat and easy to identify;
4. Keep the page clean and not contaminated by dirt, sewage and engine oil;
5. Sentences are fluent, accurate, and focused;
6. Use international units or agreed units;
7. The original record should be completed synchronously during the construction process, and no supplementary or post-recording is allowed;
8. When a major event occurs, a complete record must be made, including the cause, process, development, treatment methods, and results.

Appendix A

A.1 Terminology of Pipe Rehabilitation

A.1.1 Pipes

(1) **Conduit, pipeline, duct**: Any closed channel of any length used for conveying medium such as liquid, gas (steam), fine particles solids or for installation of pipes, cables or other facilities.

(2) **Pipeline (conduit) structure**: A general term for the hollow body structure formed by the closed channels and their ancillary facilities (pipeline accessories and ancillary structures). The channels are used for conveying various media or installation of pipelines, cables and other facilities.

(3) **Buried conduit, underground pipeline**: An underground pipe is a pipe laid underground for conveying liquid, gas, or fine particles solids.

(4) **Above-ground conduit (pipeline)**: A pipe laid directly above the ground or on a ground pier.

(5) **Submerged pipeline, subaqueous pipeline**: A pipe laid under water or in the underwater soil.

(6) **Submarine pipeline**: A pipe laid in seawater below the sea surface or on the seabed.

(7) **Overhead pipeline**: A pipeline erected above the ground, consisting of a spanning structure and a supporting structure (bracket, bracket, etc.).

(8) **Pipe bridge**: The dedicated structures which is used to build the pipeline in the form of Bridges to across the obstacles such as rivers, lakes, sea areas, railways, highways, valleys.

(9) **Industrial pipeline**: A general term for all tubular facilities which is within or between the industrial (petroleum, chemical, light industry, pharmaceutical, mining, etc.) enterprises used for the production or transmission of media.

(10) **Water supply conduit (pipeline)**: A general term for pipes conveying raw water or finished water.

(11) **Water transmission conduit (pipeline)**: Generally, it refers to the pipeline which is conveying raw water and has a certain length.

© China Architecture & Building Press 2021
L. Wang et al., *Technology Standard of Pipe Rehabilitation*,
https://doi.org/10.1007/978-981-33-4984-1

(12) **Water distribution pipeline**: The pipeline which is conveying finished water.

(13) **Drainage pipeline, sewer pipeline**: A system consisting of pipes, drains and associated facilities for the collection and discharge of sewage, waste water and rainwater.

(14) **Storm sewer conduit (pipeline)**: The pipeline which used to transport urban intercepted rainwater.

(15) **Combined drainage conduit (pipeline)**: The pipes that carry rainwater interception, domestic sewage, industrial wastewater and other combined discharge in cities and towns.

(16) **Sewage conduit (pipeline)**: The pipeline conveying treated or untreated domestic sewage or industrial wastewater from towns or industrial and mining enterprises.

(17) **Culvert**: A general term for a drainage pipe structure that crosses an embankment or river bank for the purpose of discharging surface water. It is generally composed of tunnel body pipe structure and inlet and outlet water entrance structure, including culvert, arch culvert, box culvert, cover culvert and other culvert structure types.

(18) **Pipeline for water supply in building**: A general term for water supply pipelines explicitly or implicitly installed inside industrial and civil buildings.

(19) **Pipeline for wastewater (sewerage) in building**: A general term for domestic sewage and industrial waste water pipes explicitly or implicitly installed inside industrial and civil buildings.

(20) **Down pipe, down spout**: An indoor or outdoor vertical drainage pipe that directs rainwater from the roof or platform of a building to an underground drainage pipe or other treatment means. There are circular, rectangular and other cross-section forms.

(21) **Cooling water pipeline**: A general term for pipes and associated installations that convey uncooled and cooled water between cooling substances and cooling devices.

(22) **Petroleum transmission pipeline**: A general term for pipelines and their ancillary facilities transporting crude oil or refined oil products from production, storage and other oil supply facilities to users.

(23) **Gas transmission pipeline**: A general term for pipelines and their ancillary facilities transporting natural gas, gas, etc. from production, storage and other gas supply facilities to users.

(24) **Heat-supply pipeline**: A general term for the pipes and their auxiliary facilities conveying heat supply medium from thermal power plants, boiler houses and other thermal sources to users, including above-ground laying, underground laying, trench laying, direct buried laying and other laying methods.

(25) **Heating pipeline**: A general term for pipes and accessories used for building heating, conveying heating medium from heat source or heating device to heat dissipation equipment.

(26) **Pipe duct**: Underground pipes for the laying and replacement of water-gas (steam) and other piping facilities. It is also the common name of the envelopment structure for laying and transporting heat supply medium pipelines

(Heating groove), including rectangular, circular, arch and other pipe structure forms.

(27) **Cable duct**: Underground piping for the laying and replacement of electrical or telecommunications cable facilities. It is also the enveloping structure of cable laying facilities, including rectangular, circular, arched and other pipe structures.

(28) **Accessible duct**: A general term for underground pipelines such as conduits and cable trenches in which a person may pass and carry out inspection and maintenance. An underground duct that allows a person to pass upright is called a walkway-duct. An underground duct where a person has to bend down to pass is called a semi-open trench.

(29) **Unpassable duct**: The cross section of the underground pipe trench can only meet the minimum clearance size requirements for laying pipelines or cables, and people can not enter.

(30) **Combined duct**: It is also called common ditch, which is used to lay water, gas (steam) pipes and cables in the section and transport more than two kinds of facilities with different purposes.

(31) **Combined duct**: See Combined piping.

(32) **Electrical conduits**: For the protection and protection of buildings (structures) internal and external buried, overhead electrical line systems through the penetration and replacement of telecommunications or power cables, including smooth sleeve, wave twisted sleeve, insulating sleeve, flame retardant and non-flame retardant and other different materials and performance types.

(33) **Non-pressure pipeline**: The liquid transported in the pipeline operates under the action of its own weight and gravity, and the highest operating liquid level of the liquid in the pipeline does not exceed the inner top of the pipeline section.

(34) **Pressure conduit/pressure pipeline**: A general term for pipes which conveying medium such as liquid and gas is operated under pressure. Generally expressed by atmospheric pressure, according to the requirements of different media and their corresponding working pressure, can be divided into low pressure, medium pressure, high pressure and other different pressure classes of pipelines.

(35) **Gravity-flow conduit (pipeline)**: The liquid transported in the pipeline operates under the action of its own weight and gravity. If the highest running water head does not exceed the inner top of the pipe section, it is a non-pressure pipe, and if the highest running water head exceeds the inner top of the pipe section, it is a pressurized gravity flow pipe, also known as artesian pipe.

(36) **Free-flow conduit (pipeline)**: See Gravity-flow conduit (pipeline).

(37) **Pipeline undercrossing**: A way in which a pipe passes under a natural or artificial obstacle, such as a river, railroad, highway, or building.

(38) **Pipeline aerial (over) crossing**: A means of piping over a natural or artificial obstacle such as a river, valley, railroad, highway, etc.

(39) **Pipeline trestle**: A truss or frame structure used to support an overhead pipeline, which is generally composed of supporting columns, hat beams and beams.

(40) **Pipeline pier**: A solid load-bearing structure used to support an overhead pipe, usually made of brick, stone or concrete.

(41) **Rigid pipe**: A circular pipe supported by external forces mainly depending on the strength of the pipe body material. The deformation under external load is very small and the failure of the pipe is due to the control of wall strength.

(42) **Flexible pipe**: Circular pipe with significant deformation under external load. The vertical load is mostly balanced by the elastic resistance generated by the soil on both sides of the pipe, and the failure of the pipe is usually caused by deformation rather than damage to the pipe wall.

(43) **Semi-flexible pipe**: A circular pipe whose deformation is sufficient to make the soil on both sides produce elastic resistance under the action of vertical heals. The elastic resistance of the soil supports the corresponding vertical load. The numerical value is determined by the ratio of the annular stiffness of the pipe to the elastic modulus of the soil. The structural calculation of pipe belongs to the category of flexible pipe, and semi-flexible pipe is equivalent to semi-rigid pipe.

(44) **Semi-rigid pipe**: See Semi-flexible pipe.

(45) **Liner**: Daub on the inner wall of the pipe or at the same time as the tube wall structure inside the tube surface layer, such as reinforced concrete pipe inner wall daub or inlaid plastic layer, steel tube and cast iron pipe inner wall daub cement mortar layer, glass fiber pipe inner wall hot round resin layer, is a part of the pipe structure.

(46) **Coating, surface layer**: Daub the outer surface layer formed on the outer wall of the pipe or at the same time with the structure of the pipe wall during pipe making. For example, the cement mortar layer sprayed on the outer wall of prestressed concrete pipe, the asphalt resin coil layer coated on the outer wall of steel pipe or cast iron pipe, and the thermogenic resin layer on the outer wall of glass fiber pipe. It's part of the pipeline structure.

(47) **Pipeline joint**: A general term for the connection form between adjacent orifices on a pipe or between a pipe and a pipe fitting. According to the requirements of its connection function, it can be divided into rigid connection, flexible connection, etc. According to the pipe structure, there are insert type, tongue and groove type, sleeve type, etc. According to the pipe material can be divided into welding, welding, bonding and so on.

(48) **Pipe fitting**: A general designation for all types of connections between circular pipe ends, such as tapered pipe, elbow, tee, four-way, reducer and pipe seal head. Generally, it is made according to standard specifications.

(49) **Pipe accessory**: A general term for special pipe unit such as pipe fitting, compensator, valve and its assembly.

(50) **Pipeline appurtenance**: A general term for installation of various facilities for controlling the conveying medium and construction for inspection and maintenance, such as various types of inspection Wells, valve Wells, inlet and outlet, etc. It is an integral part of pipeline engineering.

(51) **Thrust blocks**: A device for preventing horizontal and vertical pipe movement caused by axial forces, such as internal pressure or temperature action, on a

pressure pipe. Generally with concrete shallow building, commonly known as pipe pier, also known as fixed pier.

(52) **Corrosion preventive of pipes**: Measures to slow down or prevent pipes such as steel pipes and cast iron pipes from being corroded or deteriorated under the chemical and electrochemical action of internal and external media or due to the metabolic activity of microorganisms, such as coating corrosion, coil corrosion, electrical corrosion, etc.

(53) **Thermal insulation of pipes**: In order to reduce the heat loss of heat supply pipe or the influence of temperature change of the medium outside the pipe on the medium inside the pipe, the construction measures are set on the outer wall of the pipe, including filling type, daub type, winding type, prefabricated type and other thermal insulation structures. Pipe insulation is also called pipe insulation.

(54) **Pressure surge**: A sudden decrease or increase in the velocity of water, resulting in a sudden change in pressure.

(55) **Sewage**: Waste water that travels through sewers.

(56) **Waste water**: Fluid in sewer system.

(57) **Structure ID**: A number used to indicate a part of a sewer system.

(58) **Water dew point**: The temperature at which a gas precipitates the first drop of water under a certain pressure.

(59) **Hydrocarbon dew point**: The temperature at which the first drop of liquid hydrocarbon is released from a gas under a certain pressure.

(60) **Ring flexibility**: The ability of pipe to withstand radial deformation without losing its structural integrity.

(61) **Bedding angle**: The central Angle of the tube section corresponding to the arc of the lower axillary Angle that is in close contact with the backfilled gravel. The supporting reaction of the soil arc foundation within this range is represented by 2α. The supporting strength of the pipe structure is proportional to the base center angle.

(62) **Plastics inspection chamber**: Using plastic drainage pipe as wellbore, the well seat is made of plastic injection, molding or welding. A well-shaped structure connected to a drainage pipe for its clearance and inspection.

A.1.2 Pipeline rehabilitation

A.1.2.1 General terms

(1) **Pipe rehabilitation**: The work to deal with defects in pipes and auxiliary structures. According to the size of the project, pipeline rehabilitation is divided into small rehabilitation, major rehabilitation and rush rehabilitation; According to the treatment method, pipeline rehabilitation is divided into pipeline maintenance and pipeline rehabilitation.

(2) **Pipe maintenance**: The operation maintenance work such as checking and cleaning the underground pipeline and its auxiliary structure.

(3) **Emergency rehabilitation**: While the pipelines have a burst and serious leakage, maintenance operation was done under the condition of the pipeline still in use. It also known as emergency maintenance.

(4) **Structural rehabilitation**: A kind of rehabilitation technology which is done in the case of inner liner pipeline bearing the internal and external pressure.

(5) **Semi-structural rehabilitation**: A kind of rehabilitation technology which is done while the original pipeline bears external earth pressure, dynamic load and internal water pressure, the inner liner pipeline bears external water pressure and vacuum pressure.

(6) **Non-structural rehabilitation**: A kind of rehabilitation technology which is done while the internal and external pressure of pipeline is completely borne by the original pipeline body.

(7) **Structural rehabilitation**: A kind of rehabilitation method that the new liner pipe is independent of the original pipe structure and can bear external hydrostatic pressure, earth pressure and dynamic load.

(8) **Outside pipe rehabilitation**: The pipeline structure reinforcement work, leakage sealing work and other repair work which are done outside the pipeline.

(9) **Grouting**: A rehabilitation method in which the slurry (chemical slurry or cement slurry) or resin is injected into the crack area of the pipeline under the action of pressure to achieve the purpose of leakage prevention and closure. The grouting method has two kinds: grouting inside the pipeline and grouting outside the pipeline from the formation.

(10) **Grouting pressure**: Maximum pump pressure when injecting mud.

(11) **Whole renovation**: Renovation for the whole construction section of the pipeline. Reinforcement and renovate the existing piping integrally which are between two (or more) inspection Wells. This technology regulation includes ultraviolet curing, hot water curing, short pipe lining.

(12) **Localized renovation**: A renovation method for the localized leakage, damage, corrosion and collapse of the original pipe with a certain length.

(13) **Spot renovation**: Renovation for the local damage, interface dislocation, local corrosion and other defects in the old pipes, such as stainless-steel sleeve, spot CIPP, rubber stainless steel extension method, etc.

(14) **Live renovation**: Renovation for the pipe under the condition of not inter-rupting the operation function, also known as renovation without stopping production, online renovation.

(15) **Robot**: Remote control equipment with closed circuit television monitors which is mainly used for localized renovation work, such as cutting obstruc-tions inside the pipe, opening the connecting holes of the branch pipe, polishing and refilling defective areas, and injecting resin into cracks and holes.

(16) **Bypass pumping**: A pumping method is used temporarily to control or change the flow of a pipe for the purpose of cooperating with the construction.

(17) **Bypass**: A temporary fluid transfer setup to share the overflow of a sewage system.

(18) **Diverting**: A method of diverting a normal sewage flow through a special sewage system usually involves bypass pumping.

(19) **Upsizing**: Method of pipe replacement for increasing the cross-section size of old pipes

(20) **Lining**: A renovation method for pipe defects, improving its performance and prolonging its service life by laying a new pipe in the old pipe or installing lining layer in the inner wall of the old pipe.

(21) **Annular space**: A circular space between the inner wall of the original pipe and the outer wall of the lined pipe.

(22) **Chemical stabilization**: A method of pipeline rehabilitation in which one or more solutions are poured into a section of pipeline during construction so that the chemical reaction can seal the crack and improve the fluidity of pipeline.

(23) **Coating , surface layer**: Which is painted on the outer surface layer of pipelines or formed at the same time with the pipeline wall structure when the pipe is being made. Apply the outer surface layer formed on the outer wall of the pipe or at the same time with the pipe wall structure during pipe making, such as the cement mortar layer sprayed on the outer wall of the prestressed concrete pipe; Asphalt or asphalt resin coated on the outer wall of steel pipe and cast iron pipe; Thermosetting resin layer of glass fiber tube outer wall.

(24) **Liner**: The lining formed in the old pipeline through a variety of trenchless rehabilitation methods without damaging the old pipeline.

(25) **Slip lining rehabilitation**: A rehabilitation method that the new pipe is directly installed into the old pipe as the lining, and the ring gap between the old pipe and the new pipe should be filled with grouting.

(26) **Embedding rehabilitation**: A rehabilitation method that the new pipe is directly placed into the old pipe as the lining, and no ring gap fit between new and old pipes.

(27) **Lining rehabilitation**: A rehabilitation method that the liner pipe and the old pipe can form an integrated composite pipe.

(28) **Upgrading rehabilitation**: Pipeline rehabilitation to improve pipeline operating pressure, safety, etc.

(29) **Winch**: A device used to pull a CCTV camera or cleaning tool inside a sewer pipe.

(30) **Close fit**: A lining system for a tight fit between old and new pipes after renovation.

(31) **Joint sealing**: A method of inserting an expandable packer into a pipe to seal leaking joints and injecting resin or slurry to seal the joints. The packer should be recovered when the plugging is complete.

A.1.2.2 Pipe Cracking

(32) **Pipe bursting/splitting**: A pipe rehabilitation method that the original pipe is broken or cut from the inside by the crushing (cracking) pipe equipment, and the original pipe fragments are pushed into the surrounding soil to form the pipe hole, and the new pipe is pulled into the pipe hole simultaneously. This method mainly used for replacing brittle pipes such as concrete pipes and clay

pipes. There are pneumatic tube crushing method and hydraulic expansion method.

(33) **Pipe bursting, cracking**: It is a kind of pipe rehabilitation method. So as to replace the old pipe or expand the capacity, a splitter is guided by the old pipe and cut and expand the old pipe, while the new pipe is pulled in.

(34) **Burst strength**: The internal pressure required to cause pipe or casing failure.

(35) **Fracture, splitter**: A pipe breaker that breaks old pipes and squeezes pipe fragments into the surrounding soil.

(36) **Pneumatic head**: A power tool driven by compressed air that provides pulse power to drive the crusher tool forward and to break up the old pipes.

(37) **Hydraulic head**: A power tool driven by hydraulic power that provides impulse power to drive the crusher tool forward and break up the old pipe.

(38) **Static head**: A expansion head which is slightly larger than the inner diameter of the old pipe, is connected behind the drill pipe during operation, and expands the old pipe during the process of static pressure pull-back and then drags the new pipe into the old pipe.

(39) **Guide nose**: A circular wheel which is connected in front of the first pulling rod supports and protects the pulling to entry pipe.

(40) **Soil displacement**: In the process of pipe replacement, the soil around the pipe is squeezed and occurs displacements.

A.1.2.3 Sliplining

(41) **Continuous sliplining**: Also known as continuous penetration method which is placing the new liner pipe (a continuous plastic pipe) into the old pipe at one time.

(42) **Sliplining with discrete pipes**: The renovation method of inserting discontinuous short pipe in the old pipe and fixing the gap between the new pipe and the old pipe by grouting.

(43) **Short tube insertion method**: A kind of sliplining renovation method by jacking the short tube segments into the old pipe when the pit is restricted.

(44) **Pulling (dragging) method**: A method of placing liner pipes into old pipes by dragging and dropping.

(45) **Short pipe lining**: An integral pipe rehabilitation method in which the prefabricated plastic new pipe is directly inserted into the pipeline to be renovated by means of pulling and pushing, or the plastic profile or sheet material is used to join the lining, lining in the pipeline to be renovated to form a new pipe, and the gap between the new pipe and the original pipeline is filled.

(46) **Thrusting method**: A method of jacking lined pipe into an old pipe.

(47) **Dragging pit**: A work pit used for placing equipment such as a traction frame.

(48) **Interposing pit**: A Work pit for jacking equipment with PE pipe or lined pipe.

(49) **Short tube**: A joint that can be assembled into a continuous pipe in a working pit. The short pipe is usually less than 2 m steel pipe and PE pipe, which is connected as a whole in the working pit by means of welding and wire connection.

(50) **Dragging head**: In order to protect the head of the new pipe and prevent soil, water and foreign matter from entering the new pipe when pulling the new pipe into the old pipe, the conical protection joint connected in front of the new pipe has a pulling ring at the front end of the protection head, which can be connected to the winch of the working pit in front through steel cables.

(51) **Maximum allowing pulling force**: When thrust through the pipe by pulling, the maximum safe pulling force that can be used taking into account the cross-section area of the new pipeline and the tensile yield strength of the pipeline when thrust through the pipe.

(52) **Maximum allowing pulling length**: When a new tube is inserted by pulling, the new pipe drawing length is determined after taking into account the maximum allowing pulling force, mass per unit length of pipe and friction coefficient between the old and new pipes, the new pipe drawing length is determined.

(53) **Maximum allowing pushing force**: When thrust through the pipe by incremental launching method, the maximum safe jacking force that can be used taking into account the cross-section area of the new pipeline and the maximum allowable compressive stress of the pipeline.

(54) **Maximum allowing pushing length**: When a new tube is inserted by incremental launching method, the new pipe jacking length is determined after considering the maximum allowing pushing force, mass per unit length of pipe and friction coefficient between old and new pipes.

(55) **Minimum allowable radius of curvature of liner**: In the process of hoisting, winding and installation of new pipe interspersed construction, the minimum allowable turning radius is set to avoid excessive bending to damage new pipe.

(56) **Protecting connector**: An automatic disconnection device installed between the winch and the tub head during pulling intubation to prevent the winch's pull from exceeding the ultimate tensile resistance of the pipe. It is used to ensure that the tow rope and the tub head will automatically detach when the pulling force exceeds the maximum allowable traction.

(57) **Protecting ring**: In order to prevent friction damage to the outer layer of the new pipe and reduce drag resistance, short pipes are arranged at certain distance outside the new pipe.

(58) **Positioner**: When the new pipe is inserted into the old pipe, due to the dead weight of the pipe or the buoyancy of mud, the new pipe will often sink at the bottom of the old pipe or float at the top of the old pipe. In order to make the new pipe in the middle of the old pipe, the short pipe applied at intervals outside the new pipe is used to fix the new pipe in the axis position of the old pipe.

A.1.2.4 Pipe segmental lining

(59) **Segment**: It is the pipe prefabricated plate which can be assembled into a certain diameter of pipe. It includes stainless steel pipe pieces, plastic pipe pieces.

(60) **Stainless steel lining**: The renovation method of pipeline with stainless steel material as lining; For large diameter pipes that can be entered by people, the pipe billet is made of stainless-steel plate and welded manually in the pipe to form the lining layer.

(61) **PVC module assembling**: A renovation method of pipeline in which plastic modules are connected and assembled into new pipes by using bolts, and special grouting is filled between the existing pipes and the assembled plastic pipes.

(62) **Supporting segment**: In the process of assembling the pieces of pipe, the construction work which is to positioning and fixing the pieces of pipe.

(63) **Segment assembly**: Construction personnel enter into the pipe to assemble prefabricated pipe pieces into lining structure.

(64) **Splice segment lining**: A renovation method of pipeline that splice sections are spliced into a new pipe in the original pipe, and the gap between the new pipeline and the original pipeline is filled.

(65) **PVC lining concrete and reinforced concrete sewer pipe**: A pipe with PVC (modified Polyvinyl chloride) sheet lining inserted into the inner wall of the pipe during the molding process based on concrete and reinforced concrete drain pipes.

(66) **Relaining key pull strength**: A special tensile testing machine is used to measure the strength of the fixed key of PVC sheet lined with pipes, which is expressed by the tensile force on the key bar of unit length, to check whether the fixed key of PVC sheet is firmly embedded in concrete.

(67) **Sweating soldering**: After the PVC sheet overlaps with each other, the contact parts of the two connection surfaces are heated and melted together by the special hot melt tool.

(68) **Electric spark insulation performance**: High voltage electrostatic output detection equipment is used to detect the surface of PVC sheet lined with pipe line by line to test the integrity and integrity of PVC sheet inside the pipe.

A.1.2.5 Modified Sliplining

(69) **Lining with close fit pipes**: The new liner whose outer diameter is the same as the inner diameter of the old pipe shall be temporarily reduced in section size and inserted into the old pipe, and then the new lining pipe shall be restored to fit closely with the inner wall of the old pipe. This pipeline rehabilitation method does not need grouting to fill the gap between the old and new pipes. Lining with close fit pipes includes swagelining and deformed lining.

(70) **Deformed and reformed lining, swagelining**: Utilized the memory function of materials, the diameter of the circular new pipe is reduced by extrusion

or drawing. After being putted into the old pipe, the new pipe's diameter is restored to the original size by heating or pressurizing. In this way, the new 1 new pipe will fit with the old pipe closely. The method includes two technology: swage lining technology and rolldown technology.

(71) **Die shrinking-diameter method**: It relates to an interpolation lining construction method. After the polyethylene liner pipe is passed through the die compression diameter equipment forging die sleeve, the outer diameter of the line pipe is reduced and then put it into the old pipe. Using the memory function of PE material, after the external pressure is removed, the PE liner pipe is restored to its original diameter naturally or by pressure.

(72) **Roller shrinking-diameter method**: It relates to an interpolation lining construction method. After the polythene liner pipe is multi-stage rolled by the roller reducing equipment, the outer diameter of the line pipe is reduced and then put it into the old pipe. Using the memory function of PE material, after the external pressure is removed, the PE liner pipe is restored to its original diameter naturally or by pressure.

(73) **Die draw**: The new pipe is pulled out by a tapered steel die, which makes the long molecular chains of the plastic pipe reassemble, and then the pipe diameter will be reduced by 7–20%. Thus, it can be inserted into the old pipe smoothly. If the pipe diameter is small, it can be drawn at room temperature. If the pipe diameter is large, the die is usually heated to more than 100 °C ahead of the being drawn. After the liner is in place, the diameter is gradually restored to the original structure relying on the memory function of plastic molecular chain. In this way, the new line pipe can form a close match with the old pipe.

(74) **Fold and formlining , fold-and-formlining**: The method is suitable for deformable PE or PVC pipe. Using the memory function of material, the circular new pipe is folded into "U" shape or "C" shape in factory or construction site. Then the "U" shape or "C" shape pipe is pulled into the old pipe. With pressurized or heated, the "U" shape or "C" shape pipe will restore to its original shape and size. In this way, the new pipe will match with the old pipe closely.

(75) **Folded pipe**: It relates to a pipe with a U-shaped or C-shaped section which is formed by pressing and folding a circular plastic tube.

(76) **Deformed-and-reformed lining (DRP)**: A term used to describe some systems in which the liner is deformed to reduce its size during insertion, and then reverted to its original shape by the application of pressure and/or heat.

(77) **Slot**: When the thermoplastic deformed polyethylene liner is restored to its original circular shape, because of losing its supporting point at the side joint or deformation of the pipe, the liner pipe has a local deformation by expansion.

(78) **Dragging pit**: A work pit used for placing equipment such as a traction frame.

(79) **Interposing pit**: A work pit used for inserting PE pipe and installing deformed lining equipment.

(80) **Rebound**: The process relates to that the deformed tube restore to a circular shape by means of pressurization and heating after being inserted into the repaired pipe.

A.1.2.6 Cured in Place Pipe (CIPP)

(81) **Inversion method**: The hose impregnated with thermal-cured resin material is transported to the construction site. Under the action of water pressure and air pressure, the hose is inverted and putted into the pipe and close to the inner wall of the pipe. Then the resin material is cured by hot water, steam, spray or ultraviolet heating, thus forming a high-strength resin pipe inside the old pipe.

(82) **Dragging method**: The hose impregnated with thermal-cured resin material is transported to the construction site. The hose is pulled into the old pipe by traction, and then pressurized to make it expand and cling to the inside of the pipe. It's a way of curing by heating is similar with inversion method.

(83) **Inversion**: When using the CIPP to renovate the pipe, the hose is inverted and putted into the pipe with the water pressure and air pressure, making the resin impregnated side cling to the inner wall of the old pipe seamlessly, airlessly and tightly.

(84) **Resin**: Generally, it refers to the unsaturated, styrene-based thermal curing resin and catalyst system, or an epoxy resin and curing agent. The selection of resin should match the flip process adopted.

(85) **Thermosetting resin**: A kind of resin that has chemical changes after heating and gradually hardens and forms. After being hardened into shape, it does not become soft or dissolve when heated again. Thermosetting resins commonly used for new pipes include unsaturated polyester resin and epoxy resin.

(86) **Thermoplastic resin**: Resin with properties of softening by heating and hardening by cooling without having chemical reactions. No matter how many times it is heated and cooled, it can keep the properties. Thermoplastic resin commonly used for new pipes include PE and PVC and so on.

(87) **Adhesive**: A gelatinous substance used to bond a tubular composite lining material to the inner wall of a metal pipe when use inversion method to repair the pipe.

(88) **Soaking**: The process of distributing the adhesive over the hose and distributing it evenly before use inversion method to renovate the pipe.

(89) **UV protective film**: A plastic film on the outermost layer of a hose that protects the resin inside the hose's fiberglass braid from ultraviolet radiation and damage.

(90) **Inner membrane**: A plastic film in the innermost layer of the hose that protects the resin inside the hose's fiberglass braid from volatile during storage and transportation.

(91) **Soft liner, hose**: A tube consisting of one or more layers of flexible nonwoven, woven, or non-woven and woven mixed materials.

(92) **Ultraviolet light curing**: A kind of curing method relates to that the inner liner pipe is impregnated with photosensitive resin. After being placed in the old pipeline, the inner liner is cured by ultraviolet light to form the glass fiber reinforced plastic lining layer.

(93) **Hot water curing method**: An integral pipe rehabilitation method in which the inner liner impregnated with thermosetting resin is inverted or pulled into the old pipe and cured by heating water.

(94) **Steam curing method**: After the inner liner pipe is placed into the old pipe, the steam equipment distributes the steam through the pipe. The steam will make the temperature of the gas in the new pipe rise above the curing temperature of the resin. In this way, the resin of the inner liner pipe is cured.

(95) **Ambient curing method**: A kind of curing method that the resin of the inner pipe is cured under the room temperature.

(96) **CIPP of UV curing**: Without changing the location of the old pipeline, the hose soaked in resin is placed into the old pipeline by pulling or by air compress and close to the inner wall of the pipeline. Then utilizing the properties of the resin that can be cured when exposed to ultraviolet light, put the ultraviolet lamp into the inflatable hose and control the ultraviolet lamp to walk in the hose at a certain speed, it can make the hose solidify gradually from one end to the other and close to the inner wall of the pipeline to be repaired.

(97) **Roller**: The device is installed at the surface wellhead or downhole pipe mouth to change the direction of the hose being pulled, avoid the friction between the hose and the wellhead or the pipe mouth, protect the ultraviolet film of the hose, and guide the direction of the hose pulling.

(98) **Curing entry shaft**: When the CIPP of UV curing method is used to renovate the drainage pipe, a well in which a hose installed in a section of the pipeline to be repaired begins to solidify. No curing equipment is placed near the ground of curing entry shaft and the UV lamp is normally placed into the pipe from this well.

(99) **Curing reception shaft**: When the CIPP of UV curing method is used to renovate the drainage pipe, a well in which a hose installed in a section of the pipeline to be renovated is finally solidified. Compared with the curing entry shaft, the site near the ground of the curing reception shaft is larger. The curing equipment is placed at the curing reception shaft and the ultraviolet lamp is taken out from here.

(100) **Clinging**: The condition in which the outer surface of the lined pipe is in close contact with the inner surface of the old pipe after the restoration of roundness and relaxation.

(101) **Clinging liner**: A kind of lined pipe construction that the outer surface of the newly formed liner pipe is in close contact with the inner surface of the old pipe after the pipe is repaired.

(102) **Swell**: The area in which the liner pipe body is separated from the original pipe wall and formed a bulge.

(103) **Curing pressure**: The minimum air, water or steam pressure required to solidify new pipe.

(104) **Curing time**: The minimum time required to maintain curing pressure and temperature when new pipe is solidified.

(105) **Wrinkling**: Due to the inner diameter difference between the new liner pipe and the old one, or the insufficient air pressure, etc., the new pipe could not be fully expanded, resulting in the local surface undulation after solidification.

(106) **Pinhole**: Because the lining hose and impermeable film have suffered cone piercing during transportation or construction, there are holes in the solidified new pipe wall.

(107) **Notch**: The lining hose and impermeable film have been damaged in the process of transportation and construction, resulting in that the new pipe has overall discontinuity and gap defect after solidification.

(108) **Bubbling**: During the curing process of the new pipe, because the curing temperature is too high, or the adhesion between the impermeable film and the woven cloth is not strong, it is leading to the local protruding on the surface of the cured pipe.

(109) **Weak zone**: During the curing process of the new pipe, because the curing temperature is too low, or the heating time is too short or the cooling of the groundwater outside the pipe, some areas of the cured pipe where the strength and stiffness do not meet the structural strength requirements.

A.1.2.7 Spirally Wound Lining

(110) **Spiral winding**: A device that could make the plastic tape with a rebar become a lining layer on the inside of an old pipe. The ring gap between the liner and the old pipe usually needs to be fixed by grouting.

(111) **Mechanical spiral winding**: A kind of pipe renovation method that the strip profile is winded into a new pipe in the old pipe by the spiral winding machine and the gap between the new pipeline and the old pipeline is treated by grouting.

(112) **Pushing machine winding**: The winding machine is placed in the working pit. The main and secondary locks on both sides of the profile are interlocked respectively. By spiral rotation, the profile forms a new continuous seamless waterproof pipe with fixed diameter and is pushed into the old pipe by the winding machine.

(113) **Self running winding**: The winding machine moves along the old pipe, spinning and winding the pipe while the new pipe stays still.

(114) **Manual spiral winding**: For large diameter pipelines that allow people to enter, a new pipeline could be wound manually in the old pipeline and the annular space between the old and new pipelines should be grouting.

(115) **PVC lining belt**: The PVC lining belt is prefabricated profile, which is wound on site, and its shape completely conforms to the shape of the old pipe or conveying pipe.

(116) **PVC sealing connecting strip**: It is the matching product of the lining belt, and its construction principle is equivalent to producing a locking mechanism at the edge of the contour forming lining belt.

(117) **Lining strip reel**: The PVC lining tape is provided by coil, which should be placed above the working well. The end of the section should be found from the coil and pulled out, and the lining tape should be installed inside the pipe through the working well.

(118) **Section height**: The section height of the profile belt.

(119) **Section width**: The section width of the profile belt.

A.1.2.8 Spray Lining

(120) **Anti-corrosion repair**: On the premise that the original structure of the old pipeline can still bear internal pressure, external earth pressure and dynamic load, a spray repair method is used to solve the problem of corrosion and improve the roughness of pipe inner wall.

(121) **Reinforced repair**: Due to cracks, local mild damage, local corrosion and other defects on the pipe to be repaired, the original structure of the pipeline cannot fully bear internal pressure, external earth pressure and dynamic load. A spray repair method is used to make the pipe can bear internal pressure, external earth pressure and dynamic load independently.

(122) **Working pit for spray**: A work pit that is used to store the waste removed from the inner wall of the pipeline, where the nozzle begins (away from the spraying machine) or ends (close to the spraying machine) the spraying process.

(123) **Gunite**: A method of pipe rehabilitation in which steel bars are placed on the walls of an old sewage pipe and is sprayed with concrete to form a covering.

(124) **Substrate**: The inner surface of the pipe to be repaired.

(125) **Coating**: The film layer formed by curing the coating on the substrate. When the pipe is repaired by reinforced spray method, the coating is an integral new pipe.

(126) **Cement mortar coating**: In order to prevent the pipe inner wall from corrosion and scaling which will reduce the roughness of the pipe wall, the cement paste lining layer is applied on the inner wall of the pipe. There are some methods, such as mechanical spraying, hand daub and centrifugal daub.

(127) **Resin coating**: The inner wall of steel pipe and cast-iron pipe is sprayed with resin by rotary jet.

(128) **Coating thickness**: The distance between the coating surface and the substrate surface.

(129) **Wet-film thickness**: The wet coating thickness measured immediately after coating is applied.

(130) **Dry-film thickness**: The thickness of the remain coating on a surface after hardening.

(131) **Injection machine**: Machine used for pipe spraying repair operation.

(132) **Umbilical tube**: A composite pipe consist of multiple pipes which is serving for the spraying operations. Inside the composite pipe, there are two paint pipes, one compressed air pipe, three accompanying heating pipes, one accompanying wire rope. Outside the composite pipe, it is a rubber pipe. The umbilical tube can be reeled to the coil of the spraying machine.

(133) **Rotating sprayer**: A high-speed rotating nozzle used to spray the lining material (cement slurry, resin, etc.) during spraying and repairing operation.

(134) **Rotating speed**: Turns number of spray nozzle per unit time during spraying and repair.

(135) **Walking speed**: The travel distance of the jet along the pipe axis in unit time during spraying repair operation.

(136) **Jet velocity**: The speed at which the lining material is ejected by the rotating nozzle during the spray repair operation.

A.1.2.9 Localized Renovation

(137) **Resin injection**: A localized renovation method for pipe which is often used in sewage pipes. By injecting resin into cracks and holes, it can prevent leakage and further deterioration after curing, and can also increase the structural strength.

(138) **Stainless-steel lining**: A stainless steel sleeve with adsorbed styrofoam sponge is installed at the partial damage of the pipe. After the styrofoam expands, a sealable contact is formed between the old pipe and the stainless-steel sleeve.

(139) **Localized repair**: Methods to repair the local damage, interface dislocation, local corrosion and other defects in the original pipeline. In this regulation, it means spot cured-in-place pipe method and stainless-steel sleeve method.

(140) **Spot CIPP, spot cured-in-place pipe**: A localized renovation method of applying resin-impregnated fabric to the damaged part of the pipe by balloon dilation and then curing the pipe by heating.

(141) **Joint sealing**: A localized renovation method of inserting an expandable packer into a pipe and injecting resin or slurry to seal the joints.

(142) **Plugging lining with hoop**: A localized renovation method for pipe leakage by putting a pipe hoop outside of the pipe wall.

(143) **Fracture embedment method**: In the renovated position of original reinforced concrete pipe, water-soluble polyurethane slurry is grouting into the pipe with a certain grouting technology to fill and close the defective parts, which make the pipe impermeability and reinforcement.

(144) **Stainless steel double expansion ring method**: With the annular rubber sealing belt and stainless steel sleeve ring as the main repair materials, double expansion ring of rubber ring is installed at the pipe interface or local damaged part. After the double expansion ring of rubber ring is in place, it is fixed with 2–3 stainless steel expansion rings to achieve the purpose of no leakage.

A.1.3 Subsidiary work

A.1.3.1 Pipeline detection

(1) **General survey of underground pipeline**: A process that adopt an economical and reasonable method to find out the underground pipeline condition in the region, obtain accurate data about the pipeline, compile the pipeline diagram, establish database and information management system, and implement dynamic computer management of pipeline information.

(2) **Actuality survey and drawing**: All professional pipeline ownership units were responsible for organizing relevant professionals to collect data on the buried underground pipelines, and classify, organize, and prepare status modulation drawing, so as to provide reference and relevant underground pipeline attribute basis for field detection operations.

(3) **Underground pipeline information system**: With the support of computer software, hardware, database and network, adopting GIS technology, computer management system can input, edit, storage, query statistics, analyze, maintain, renovate and output of the space and attribute information of underground pipelines and affiliated facilities.

(4) **Surveying point of underground pipeline**: During the exploration of underground pipelines, in order to accurately describe the characteristics of underground pipelines and the information of auxiliary facilities, the detection points are set up in the exploration and investigation of underground pipelines.

(5) **GPSRTK (global positioning system really time Kinematic)**: Global satellite positioning system real-time differential positioning measurement method.

(6) **Locating wire**: A metal wire used to mark the position of a pipe. It is located above a pipe and can be detected from the ground by a dedicated device.

(7) **Locating tape**: A metal tape used to mark the position of a pipe. It is located above a pipe and can be detected from the ground by a dedicated device.

(8) **Ground penetration, probing radar, GPR**: Using pulse radar system, continuous firing pulse width to the underground video pulse for a few nanoseconds, accept reflected electromagnetic pulse signal, can be used to detect underground metal or nonmetal goals.

(9) **Radar detector**: An instrument that USES electromagnetic waves to detect the position of an underground target.

(10) **Pipeline locator, detector**: An instrument that use geophysical exploration method to detect the attribute and spatial location of underground pipeline.

(11) **Leak detector**: An instrument that use the acoustic principle to detect the underground pipe leakage.

(12) **Vacuum excavator**: An equipment that use high pressure water or compressed air to cut loose soil, at the same time through the soil suction vacuum suction going, quickly excavated a hole in the ground equipment.

(13) **Potholing**: A exploring underground pipelines method by forming a hole in the ground with vacuum suction.

A.1.3.2 Pipeline cleaning

(14) **Pigging system**: Pigging system is a complete set of equipment use to remove the condensates and sediments in the pipe, isolate, replace or inspect the pipe online. The commonly used pipe cleaning methods is mechanical cleaning, hydraulic cleaning and chemical cleaning.

(15) **Mechanical cleaning**: The inner wall of the pipe is cleaned by friction descaling using tools such as bar, barrel and capstan brush.

(16) **Hydraulic cleaning**: The inner wall of the pipe is cleaned by high-pressure water or water flow with head pressure. Hydraulic washing equipment includes high-velocity jet head, cleaning ball and hinge disc cleaner.

(17) **Chemical cleaning**: It is a pipe cleaning method with injecting the chemical agent into the pipeline, making the chemical cleaning solution flow in the pipeline, and then the chemical agent and the scale formation on the inner wall of the pipe will have chemical reaction. It can accelerate the stripping and cleaning the pipe.

(18) **PIG**: A bullet-shaped cleaning tool made from a special polyurethane foaming system. The commonly used pigging tools are spherical pigging (PIG ball), cup-type pigging and soft pigging, which can be equipped with instruments for measuring tube wall thickness, internal corrosion, pipe deformation and position settlement.

(19) **Pigging**: Using the pressure of the fluid in the cleaned pipe or the water pressure or air pressure provided by other equipment as the power to push the pig forward in the pipe, scrape the dirt on the pipe wall, and discharge the dirt and clean out debris accumulated in the pipe.

(20) **Sandblasting cleaning**: A method of cleaning and derusting by using compressed air as power to form a high-speed jet to spray the materials (copper ore, quartz sand, emery sand, iron ore) to the inner surface of the pipeline at high speed.

(21) **Preparatory cleaning**: Before inspecting the pipeline, water jet is usually used to remove debris inside the pipeline.

(22) **Obstacle removing with robot**: The robot is guided by the internal video camera system to locate and remove garbage and foreign matter in the pipe.

(23) **Pushrod cleaning**: A method of using manpower to push bamboo, steel bars, ditch sticks and other tools into pipes to clear blockages.

(24) **Winch bucket cleaning**: The method of dredging pipe by drawing dredging device inside pipe with winch.

(25) **Cleaning bucket**: Bucket, shovel and other tools used in dredging winch.

(26) **Purging**: The removal of residual gas and dirt from a gas facility prior to commissioning or maintenance.

(27) **Relief**: The process of using the dispersion equipment to empty the air, gas or mixed gas in the gas facilities.

(28) **Hot-tapping**: Use special machines and tools to process holes in pressurized gas pipelines and operate without gas leakage.

(29) **Plugging**: The plugging head is pulled into the pipe from the opening and sealed the pipe to prevent the flow of medium in the pipe.

A.1.3.3 Pipeline inspection

(30) **In-line inspection**: The inspection performed by an inspector running in the pipeline that can collect and record pipeline information in real time. It is also called online inspection.

(31) **Geometry inspection**: It is an in-pipe inspection designed to inspect the geometry deformation of the pipeline.

(32) **Quick inspection**: It is a peep-specific quantitative inspection method which is conducted by adopting special laser generator, image measurement and evaluation software and closed-circuit television system.

(33) **Pipe quick view (QV.)**: The device that checks the state of the pipe at the bottom of the well by a camera on a long rod.

(34) **Pipe quick view inspection**: A test method of placing a pipe periscope (a high magnification camera) into the well or concealed space by joysticks and showing the internal cracks and blockages clearly.

(35) **Closed circuit television (CCTV) inspection**: Use the professional closed circuit television camera equipment to collect and transmit images of the internal defects and conditions of the pipeline, and use the professional software to analyze the image data in order to understand and evaluate the internal conditions of the pipeline.

(36) **Sonar inspection**: An in-pipeline inspection method that uses the high-frequency acoustic signal transmitter and water as the medium to scan the inner wall of the pipeline, distinguish the strength of the acoustic signal by color, and judge the scale formation and deposition status of the inner wall of the pipeline by special detection and analysis software.

(37) **Operation inspection**: Test the degree of unblocked pipe.

(38) **Structure inspection**: Inspection of pipe construction soundness.

(39) **Geometry pigging**: The pipe inspection aims to inspect the geometric deformation of pipelines.

(40) **Corrosion pigging**: The pipe inspection aims to inspect the metal loss in the pipe wall.

(41) **Internal metal loss**: The mental loss occurs on the inner surface of the pipe wall and inside the pipe body.

(42) **External metal loss**: The metal loss occurs on the outer surface of the tube wall.

(43) **Mental loss inspection**: An in-pipe test for the purpose of inspecting metal loss in the pipe wall.

(44) **Gauge plate**: A round soft metal plate (usually made of aluminum) is mounted on a PIG and its diameter is smaller than the minimum inner diameter of a normal pipe. It used to inspect diameter changes of the pipe.

(45) **Clock description**: Use the clock position to describe the position of the defects or structural features appearing in the annular position of pipelines.

(46) **Downstream**: The direction of pipeline inspection is consistent with the direction of water flow.

(47) **Upstream**: The direction of pipeline inspection is opposite to the direction of water flow.

(48) **Above-ground marker (AGM)**: A portable or permanent device placed above a pipe is capable of inspecting and recording the passing signal of an internal inspector or inspecting and recording its emitted signal.

(49) **Buckle**: A pipe subjected to a large plastic deformation, resulting in permanent creasing or deformation of the wall or section of the pipe.

(50) **Corrosion**: The deterioration of a material (usually metal) as a result of a chemical or electrochemical reaction with its environment.

(51) **Geographical information system (GIS)**: A computer system which is capable of integrating, storing, processing, and displaying geographic reference information.

(52) **Geometry tool**: A measuring inspector that records the geometry of a pipe or wall.

(53) **Magnetic particle inspection (MPI)**: A kind of nondestructive testing technique for locating surface cracks in steel materials using fine magnetic powder and magnetic field.

(54) **Standard resolution tool**: Coil sensors with large sensor size and spacing are usually used to grade the depth of defects, such as mild (0–30% wall thickness), moderate (30–50% wall thickness), severe (50% wall thickness), generally it cannot distinguish internal defects or external defects.

(55) **High resolution tool**: The principle is basically the same as the standard resolution detector. Hall sensor is usually used, but the size, number and spacing of the sensors are smaller (usually 0–7 mm), which can accurately quantify parameters such as the length and depth of defects and distinguish internal defects from external defects.

(56) **Extra high resolution tool**: Compared with the high resolution of the detector, its sensor size is smaller, the sensor number and the spacing between sensor are smaller (usually 4–8 mm). It usually collect two (axial and radial) or two or three axial, radial and ring (to) the direction of the data, can be very precise quantitative defect parameters such as length and depth. It is not only able to distinguish between internal defects or external defects, will also be able to detect the sag, and other defects.

(57) **Pipeline feature**: Metal loss on the pipe wall, sag, elliptic deformation, auxiliary facilities of the pipe and manufacturing defects existing on the pipe.

A.1.3.4 Pipeline evaluation

(58) **Sewers canner and evaluation technologies set**: Like closed circuit television (CCTV), it can provide the front screen, can also provide 360 DHS scanning the visual image of the pipe inner surface, can analyze data in the office later, promise not to ignore some important pipeline defect assessment technology. This system can also record pipe slope, so the available pipes prolapse and

sediment of potential location, 360 DHS scanning plane view to detect the entire surface of the pipe, but also can measure the joint gap.

(59) **Renovation index**: A value is calculated according to the structural defect's degree and number of pipelines by a certain formula. The higher the value is, the greater the intensity of repair.

(60) **Maintenance index**: A value is calculated according to the type, score, number and influencing factors of pipeline functional defects. The higher the value is, the greater the urgency of pipeline maintenance.

(61) **Structural defect**: Defects affecting the strength, stiffness and tightness of the pipeline, such as deformation, fracture, wrong opening, leakage and other structural defects which can only be eliminated by semi-structural repair or structural repair means.

(62) **Functional defect**: The pipe structure is not damaged and its defects only affects the overcurrent capacity or water quality defects.

(63) **Roughness coefficient**: It is used to show the effect coefficient of the head loss caused by the rough pipe.

(64) **Partially deteriorated pipe**: The pipe has slight structural damage, but the pipe can still bear external earth pressure and dynamic load within its design life, or its section deformation is not greater than 2.5% of the pipe nominal inner diameter.

(65) **Fully deteriorated pipe**: The pipe has serious structural damage. The pipe cannot still bear external earth pressure and dynamic load although it is still within its design life, or its section deformation is greater than 2.5% of the pipe nominal inner diameter.

(66) **Corrosion**: Physical and chemical interactions between metals and environmental media, which result in changes in the properties of metals and often result in damage to the metals, the environment or the functions of the technical systems.

(67) **Biological corrosion**: A corrosion of piping material caused by the action of organisms (e.g. bacteria, algae, fungi).

(68) **Corrosion rate**: The amount of mass lost per unit time due to corrosion.

(69) **Crack, cracking**: An axial or radial crack in a pipe characterized by a sharp tip and a large aspect ratio of cracking displacements.

(70) **Deformation**: Permanent changes in shape, such as bending, buckling, depression, ellipticity, ripple, fold, or other changes that affect roundness or flatness of pipe section.

(71) **Dent**: Local plastic deformation in which the surface curvature of a pipe changes significantly due to impact or extrusion by external forces.

(72) **Spring back**: After the removal of external constraints, the sag depth of the pipeline decreases due to elastic unloading.

(73) **Rebounding**: The decrease of sag depth under internal pressure.

(74) **Dent depth**: The maximum reduction in pipe diameter relative to the original diameter at the concave position.

(75) **Collapse**: Failure of a pipeline due to structural fracture.

(76) **Encrustation**: Saline groundwater seeps into the tube, after its evaporation, it left behind deposits in the pipe.

(77) **Erosion**: Pipe surface deterioration caused by wear and tear of fluids.

(78) **Pitting**: Some local deep punctured erosion caused by high corrosion in the pipe.

(79) **Infiltration**: The phenomenon that clean water, rainwater or ground water enter the pipe through cracks, defective joints, man-hole or maintenance well, or the phenomenon that the medium inside the pipe flows to the outside.

(80) **Spalling**: The phenomenon that part of the pipeline structure is separated from the pipeline parent.

(81) **Scaling , tuberculation**: A tuberculous mass resulting from localized corrosion at different sites.

(82) **Sediment**: The deposition of small particles in the pipe can result in a reduction in cross-sectional area.

(83) **Segment**: A section of pipe that serves as an evaluation unit.

(84) **Section**: Each part of a pipeline divided for the purpose of risk assessment of the pipeline. It is the smallest unit of risk assessment of the pipeline.

(85) **Failure consequence**: The degree of adverse effects caused by the leakage accident in the pipe section, such as casualties and environmental losses.

(86) **Attribute**: Description of relevant characteristics of pipelines, including characteristics of pipeline ontology, pipeline operation, and surrounding environment.

(87) **Semi-quantitative risk assessment method**: It is a kind of pipeline risk assessment method. It establishes an index system according to pipeline attributes and their contribution to risk, scores the failure possibility and failure consequence of each pipeline segment, and uses the score value to represent the pipeline risk.

(88) **Semi-quantity risk assessment**: The process of scoring various factors that affect the possibility and consequence of failure according to the scoring system and synthesizing the risk value represented by the score.

(89) **Failure probability**: The probability of a pipeline accident which is expressed in fraction.

(90) **Basic model for failure probability assessment**: The unified failure probability scoring model is adopted in the evaluation of pipeline failure probability. This model includes the basic factors affecting pipeline failure, namely, scoring items, and gives the default weight of these scoring items.

(91) **Third-party damage**: Unintentional damage to the pipeline caused by a person other than the owner, the user, management and maintenance party of the pipeline.

(92) **Pipeline essential safety quality**: The safety level of pipeline determined by factors such as pipeline design, manufacturing, construction, geological conditions, natural disasters, defects, etc.

(93) **Risk value**: The product of the probability of failure and the consequences of failure which is expressed in fractions.

A.1.3.5 Pipeline test

(94) **Leakage test**: Take gas as medium, the test is done to detect the leakage point in the pipeline system by using blowing agent, color developing agent, gas molecular sensor or other means under the design pressure.

(95) **Water pressure test**: Take water as the medium, make the water full of the laid pressure pipeline, test whether the pipeline structure is damaged and whether its water percolating capacity conforms to the stipulated standard of allowable seepage water (or allowable pressure drop) at the specified pressure value. This test method is mainly used for the pressure pipeline conveying liquid.

(96) **Closed water test**: Test to check whether the laid pipe segment conforms to the specified allowable leakage standard by water injection method according to the specified head, generally used for gravity flow pipeline (non-pressure pipeline).

(97) **Leak test**: It is also called functional test to inspect the leakage of pipelines with liquid or gas in laid pipelines. There are two kinds of tests: closed water test and closed gas test.

(98) **Closed water test**: Tests to check whether the laid pipe segment conforms to the specified allowable leakage standard by water injection method according to the specified head, generally used for gravity flow pipeline (non-pressure pipeline).

(99) **Closed air test**: Tests on laid pipe sections to check whether they conform to the specified leakage amount at the specified pressure value by means of inflation, mainly used in pipelines conveying gases and inflammable, explosive or toxic media.

A.1.3.6 Construction management

(100) **Construction planning**: A work plan prepared by the construction unit and approved by the competent department for the completion of the trenchless project work objectives.

(101) **Management of construction production**: In order to ensure the coordination and continuity during the trenchless construction production process, a set of management work is done including planning, organizing, directing, controlling and regulating.

(102) **Pipeline design**: The design and planning department puts forward the principle of the general outline of pipeline laying track, and the competent unit or construction unit designs the pipeline track according to the trenchless construction process characteristics, pipeline conditions, and the conditions around the construction site. The optimized pipeline parameters and construction methods shall be submitted to the design and planning department for approval.

(103) **Construction design**: With the requirements of pipeline design, the construction unit shall design the trenchless construction scheme according to the pipeline track, engineering geological conditions and pipeline parameters.

(104) **Construction planning**: The overall arrangement of project construction compiled by the construction unit. It is the basis for making production plan and organizing production.

(105) **Dispatching and scheduling of construction job**: A plan prepared to coordinate production and support departments to ensure the completion of production tasks.

(106) **Norm management**: To assess the completion of production tasks and their economic benefits according to production labor quota and materialized labor quota.

(107) **Management of equipment**: The management work is done to ensure the integrity of construction equipment, reasonable allocation, reasonable use, maintenance and regular overhaul.

(108) **Technical management**: Management work of construction design, technical breakthrough, technical promotion, technical training and implementation of regulations and standards.

(109) **Technical file**: Special documents established by the construction entity to record the technical and economic activities of construction and production.

(110) **Engineering quality inspecting rule**: A system for organizing overall quality inspection of the project after completion of construction.

(111) **Installation acceptance rule**: The system for acceptance of construction equipment after it is installed and checked by relevant personnel.

(112) **Post responsibility rule**: During construction, the full-time and responsible post system shall be established according to the labor division of staff.

(113) **Environmental construction**: Use non-toxic, non-(low) pollution materials that meet environmental protection standards; Disposal and discharge of engineering muck and slurry in accordance with environmental protection requirements. The noise and traffic influence of the works should be controlled within the allowable limit.

(114) **Direct safety of constructing**: Prevent personal accidents during construction. Prevent damage to construction equipment, instruments and pipes during construction. Prevent each link technical inappropriateness to cause the project to scrap.

(115) **Indirect escape risk**: Avoid damage to the original underground pipelines and buildings underground or unground in the construction area. Prevent ground from uplift, subsidence and riser.

(116) **Engineering safety accident**: The unexpected emergencies occurs in the process of engineering construction which usually cause casualties or heavy property losses, result in the normal construction activities interrupted.

(117) **Engineering quality accident**: Due to unqualified or defective project quality, events that cause certain economic losses, delays in construction or endanger the safety of human life and the normal order of society.

(118) **Limited space**: A space which is enclosed or partially enclosed, or its import and export is narrow and limited. It is not designed as a fixed workplace, poor natural ventilation, prone to toxic and harmful, flammable and explosive material accumulation or insufficient oxygen.

(119) **Limited space work**: Operators enter the limited space to carry out activities.

A.1.3.7 Quality inspection and acceptance

(120) **Acceptance**: On the basis of the construction unit's own quality inspection and assessment, a kind of construction test activity which is organized by the project quality acceptance of responsibility (supervision or the construction unit) and with attendance of all the relevant units to confirm division (a branch), unit (units) engineering quality whether achieve qualified or not by sampling the reinspection, audit of construction materials, and according to the contract documents, design documents and relevant standards in written form.

(121) **Check (inspection)**: The activity of measuring, checking and testing the quality parameters or quality characteristics of a project or project material inspection project and comparing the results with the requirements specified in the standard to determine whether the project or project material is qualified.

(122) **Site acceptance**: Inspect the materials, components and equipment entering the construction site according to the contract, design requirements and relevant technical standards, and confirm whether the products are qualified or not.

(123) **Check (inspection) lot**: An inspection body consisting of a certain number of samples, aggregated for inspection under the same production conditions or in a prescribed manner.

(124) **Evidential testing**: In the construction unit or the construction supervising units under the witness of the construction unit of the field test of personnel involved in the engineering structure safety test blocks, test samples and materials on site, and through the provinces (municipalities directly under the central government) to the construction administrative department at or above the level of their recognition of qualifications and the metrological certification of quality and technical supervision department quality testing unit for testing.

(125) **Handing over check (inspection)**: Activities to be inspected by both the undertaking party and the completion party and to confirm whether the work can continue.

(126) **Dominant item**: Tests in engineering that are decisive for safety, health, the environment and the public interest.

(127) **General item**: Inspection items other than dominant items.

(128) **Sampling check (inspection)**: According to the specified sampling scheme (frequency), randomly select a certain number of samples from incoming materials, components, equipment or engineering inspection items according to the inspection batch.

(129) **Sampling frequency**: The sampling range and the number of points to be drawn according to the characteristics of the items to be tested.

(130) **Qualified rate**: The percentage of the number of points qualified in the same inspection item and the number of points to be inspected in the same inspection item.

(131) **Quality of appearance**: A qualitative assessment of the external quality of a project made by the project quality inspector through visual observation and necessary measurement means (measured and measured).

(132) **Rework**: The replacement, remaking and construction of the unqualified parts shall be taken.

(133) **Peels strength**: A unit of force N/mm required to strip a unit width adhesive tape from a particular adhesive surface at a given Angle and rate.

(134) **Test assembly**: A component formed by gluing the test tape to a steel plate.

(135) **Roughness average**: The arithmetic mean deviation of the roughness contour is the arithmetic mean of the absolute deviation of the roughness contour within the sampling length.

A.2　Pipe Defect Grade Classification and Sample Figures

A.2.1 Structural defect

A.2.1.1 Fracture

Defect name: Fracture			Defect code: PL
Definition: The external pressure of the pipe exceeds its own capacity and causes the pipe to break. There are longitudinal, annulus and compound three kinds of fractures			
Grade	Definition	Value	Sample figure
1	Cracks: When one or more of the following situations exist: (1) A fine crack can be seen on the wall of the pipe (2) A small amount of sediment emerges from a fine crack in the wall of the pipe (3) Mild peeling	0.5	

(continued)

(continued)

Defect name: Fracture	Defect code: PL

Definition: The external pressure of the pipe exceeds its own capacity and causes the pipe to break. There are longitudinal, annulus and compound three kinds of fractures

Grade	Definition	Value	Sample figure
2	Split: A clear gap has formed at the rupture, but the shape of the pipe is not affected and the rupture does not peel off	2	
3	Broken: At the broken or detached place of the pipe wall, the arc length of the annular coverage of the remaining debris is less than 60°	5	
4	Collapse: When one or more of the following situations exist: (1) The annular coverage of the cracks, breach and broken in the pipe wall is greater than 60° the ace length (2) The arc length of the annular range of peeling off in pipe wall material is greater than 60° (3) Deformation range of the pipe is greater than 25% of pipe diameter	10	

A.2.1.2 Deformation

Defect name: Deformation			Defect code: BX
Definition: Due to the compress of the external force, the pipe shape has changed			
Grade	Definition	Value	Sample figure
1	The deformation is less than 5% of pipe diameter	1	
2	The deformation is 5–15% of pipe diameter	2	
3	The deformation is 15–25% of pipe diameter	5	
4	The deformation is greater than 25% of pipe diameter	10	
Defect description	1. This type of defect only applies to flexible tubes 2. The confirmation of deformation percentage should be based on the actual measurement 3. $\eta = \frac{R_0 - R_{min}}{R_0} \times 100\%$ η stands for deformation rate; R_0 stands for original inner diameter; R_{min} stands for the minimum inner diameter after deformation		

A.2.1.3 Corrosion

Defect name: Corrosion		Defect code: FS	
Definition: Due to the corrosion, the inner wall of the pipe appears run off or peel off and the surface expose unsmooth or rebar			
Grade	Definition	Value	Sample figure
1	Mild corrosion: The inner surface of the pipe is slightly peeled off, and the pipe wall appears concave and convex	0.5	
2	Moderate corrosion: The inner surface of the pipe appears peel off and the coarse aggregate or rebar is partly exposed	2	
3	Severe corrosion: The coarse aggregate or rebar is fully exposed	5	

A.2.1.4 Misalignment

Defect name: Misalignment		Defect code: CK	
Definition: Two pipes at the same junction have lateral deviation and are not in the correct position. The adjacent pipes look like "half-moons"			
Grade	Definition	Value	Sample figure
1	Mild misalignment: The deviation of the two connected pipe is less than 1/2 of the pipe wall thickness	0.5	
2	Moderate misalignment: The deviation of the two connected pipe is between ½ and 1 of the pipe wall thickness	2	
3	Severe misalignment: The deviation of the two connected pipe is between 1 and 2 times of pipe wall thickness	5	
4	Most severe misalignment: The deviation of the two connected pipe is greater than 2 times of pipe wall thickness	10	

A.2.1.5 Fluctuation

Defect name: Fluctuation		Defect code: QF	
Definition: The interface position is shifted and the vertical position of the pipe has changed. Thus, the water is accumulated in the lower place			
Grade	Definition	Value	Sample figure
1	$\dfrac{\text{The height of the fluctuation}}{\text{the pipe diameter}} \le 20\%$	0.5	
2	$20\% < \dfrac{\text{The height of the fluctuation}}{\text{the pipe diameter}} \le 35\%$	2	

(continued)

(continued)

Defect name: Fluctuation	Defect code: QF	
Definition: The interface position is shifted and the vertical position of the pipe has changed. Thus, the water is accumulated in the lower place		
Grade	Definition	Value	Sample figure
3	$35\% < \dfrac{\text{The height of the fluctuation}}{\text{the pipe diameter}} \leq 50\%$	5	

(continued)

(continued)

Defect name: Fluctuation		Defect code: QF

Definition: The interface position is shifted and the vertical position of the pipe has changed. Thus, the water is accumulated in the lower place

Grade	Definition	Value	Sample figure
4	$\dfrac{\text{The height of the fluctuation}}{\text{the pipe diameter}} > 50\%$	10	

Defect description	H is the height of the fluctuation, that is, the value of the pipe deviating from the design height

A.2.1.6 Disconnection

Defect name: Disconnection		Defect code: TJ	
Definition: The ends of the two pipes are not sufficiently joined or come apart. The adjacent pipes look like "a full moon"			
Grade	Definition	Value	Sample figure
1	Mild disconnection: A small amount of soil has been forced into the pipe at the end of the pipe	1	
2	Moderate disconnection: The disconnection distance is less than 2 cm	3	
3	Severe disconnection: The disconnection distance is 2–5 cm	5	
4	Most severe disconnection: The disconnection distance is greater than 5 cm	10	
Defect description	Schematic diagram of the pipeline disconnection		

A.2.1.7 Interface material shedding

Defect name: Interface material shedding	Defect code: TL		
Definition: Rubber ring, asphalt, cement and similar interface materials fall into the pipe			
Grade	Definition	Value	Sample figure
1	In the pipe, the interface material may be visible above the horizontal center line of pipe	1	
2	In the pipe, the interface material may be visible below the horizontal center line of pipe	3	

A.2.1.8 Branch pipe connected laterally

Defect name: Branch pipe connected laterally		Defect code: AJ	
Definition: Without passing through the inspection well, the branch pipe is connected to the main pipe laterally			
Grade	Definition	Value	Sample figure
1	The length of branch pipe entered into the main pipe is less than 10% of the main pipe diameter	0.5	
2	The length of branch pipe entered into the main pipe is between 10 and 20% of the main pipe diameter	2	
3	The length of branch pipe entered into the main pipe is greater than 20% of the main pipe diameter	5	

A.2.1.9 Foreign objects penetration

Defect name: Foreign objects penetration	Defect code: CR		
Definition: Objects that don't belong to the pipe ancillary facilities penetrate the pipe wall and enter into the pipe			
Grade	Definition	Value	Sample figure
1	The foreign object occupies less than 10% of the water section in the pipe	0.5	
2	The foreign object occupies 10–30% of the water section in the pipe	2	
3	The foreign object occupies greater than 30% of the water section in the pipe	5	

A.2.1.10 Leakage

Defect name: Leakage		Defect code: SL	
Definition: Water outside the pipe flows into the pipe or water inside the pipe leaks out of the pipe			
Grade	Definition	Value	Sample figure
1	Drop leakage: Water continuously drips from the defect point of the pipe and flows along the pipe wall	0.5	
2	Line leakage: Water continuously flows from the defect point of the pipe and flows away from the pipe wall	2	
3	Floating leakage: A large amount of water gush or spray from the defect point of the pipe and the gushing leakage surface area is less than 1/3 of the pipeline section	5	
4	Spray leakage: A large amount of water gush or spray from the defect point of the pipe and the spray leakage surface area is more than 1/3 of the pipeline section	10	

A.2.2 Functional defect

A.2.2.1 Sediment

Defect name: Sediment		Defect code: CJ	
Definition: Impurities settle and accumulate at the bottom of the pipe			
Grade	Definition	Value	Sample figure
1	The sediment thickness is less than 20–30% of the pipe diameter	0.5	
2	The sediment thickness is between 30 and 40% of the pipe diameter	2	
3	The sediment thickness is between 40 and 50% of the pipe diameter	5	
4	The sediment thickness is greater than 50% of the pipe diameter	10	
Defect description	1. Use clock representation to indicate the location of sediment 2. Soft or hard should be marked 3. The sonar image should measure the maximum deposition		

A.2.2.2 Scaling

Defect name: Scaling		Defect code: JG	
Definition: An attachment on the inner wall of a pipe			
Grade	Definition	Value	Sample figure
1	The loss of water passage section caused by hard scaling is less than 15% The loss of water passage section caused by soft scaling is 15–25%	0.5	
2	The loss of water passage section caused by hard scaling is 15–25% The loss of water passage section caused by soft scaling is 25–50%	2	
3	The loss of water passage section caused by hard scaling is 25–50% The loss of water passage section caused by soft scaling is 50–80%	5	

(continued)

(continued)

Defect name: Scaling		Defect code: JG	
Definition: An attachment on the inner wall of a pipe			
Grade	Definition	Value	Sample figure
4	The loss of water passage section caused by hard scaling is more than 50% The loss of water passage section caused by soft scaling is more than 80%	10	
Defect description	1. Use a clock representation to indicate the location of scaling 2. The percentage of the cross-section loss should be calculated and indicated Soft or hard should be marked		

A.2.2.3 Obstacle

Defect name: Obstacle		Defect code: ZW	
Definition: A blockage in a pipe that affects overflow			
Grade	Definition	Value	Sample figure
1	Section loss is less than 15%	0.1	
2	Section loss is 15–25%	2	
3	Section loss is 25–50%	5	
4	Section loss is more than 50%	10	
Defect description	The type of obstacle and the section loss rate should be recorded		

A.2.2.4 Residual wall

Defect name: Residual wall		Defect code: CQ	

Definition: When the pipe is tested for closed water, temporary masonry wall is built for sealing. After the test, the remains are not removed or not thoroughly removed

Grade	Definition	Value	Sample figure
1	Section loss is less than 15%	1	
2	Section loss is 15–25%	3	
3	Section loss is 25–50%	5	
4	Section loss is more than 50%	10	

A.2.2.5 Tree root

Defect name: Tree root		Defect code: SG	
Definition: A single root, or groups of roots, grow naturally into the ducts			
Grade	Definition	Value	Sample figure
1	Section loss is less than 15%	0.5	
2	Section loss is 15–25%	2	
3	Section loss is 25–50%	5	
4	The section loss is more than 50%	10	

A.2.2.6 Floatage

Defect name: Floatage	Defect code: FZ

Definition: Floatage on the surface of the pipe. (The defect shall be recorded in the test record, not involved in calculation)

Grade	Definition	Sample figure
1	A small amount of floatage which occupy less than 30% of the water surface area	
2	Much more floatage which occupy 30–60% of the water surface area	
3	A larger number of floatage which occupy more than 60% of the water surface area	

Bibliography

1. Rubber hot air aging test method. GB3512
2. Tensile properties of textile fabrics part 1: determination of breaking strength and breaking elongation strip method. GB/T3923.1
3. Specifications for the preparation of plastic pipes and fittings polyethylene (PE) pipes/pipes or pipes/pipes hot-melt butt assemblies. GB19809
4. Determination of the performance of plastics resistant to liquid chemical reagents. GB/T11547
5. Buried polyethylene (PE) pipeline system for gas part 1: pipes. GB15558.12003
6. Head of pressure vessel. GB/T25198
7. Noise limits for construction sites. kGB12523
8. Centrifugal casting glass fiber reinforced unsaturated polyester resin sand pipe. JC/T695
9. Code for design of underground fiber reinforced plastic sand pipe structure for water supply and drainage engineering. CECS 190
10. Glass fiber reinforced plastic sand pipe. GJ/T3079
11. Glass specification for interface sealing rings for water supply, drainage and sewage of rubber seals. HG/T3091
12. Building construction safety inspection standard. KJGJ59
13. Environmental and sanitary standards for construction sites. JGJ146
14. Safety technical regulations for construction machinery. JGJ33
15. Technical regulations on safety of temporary electricity on construction site. KJGJ46
16. Code for structural design of buried steel pipeline of water supply and drainage engineering. CECS141:2002
17. Technical regulations for detection and evaluation of drainage pipeline television and sonar. DB31 T444-2009

© China Architecture & Building Press 2021
L. Wang et al., *Technology Standard of Pipe Rehabilitation*,
https://doi.org/10.1007/978-981-33-4984-1

Printed in the United States
by Baker & Taylor Publisher Services